206
Advances in Polymer Science

Editorial Board:
A. Abe · A.-C. Albertsson · R. Duncan · K. Dušek · W. H. de Jeu
J.-F. Joanny · H.-H. Kausch · S. Kobayashi · K.-S. Lee · L. Leibler
T. E. Long · I. Manners · M. Möller · O. Nuyken · E. M. Terentjev
B. Voit · G. Wegner · U. Wiesner

Advances in Polymer Science
Recently Published and Forthcoming Volumes

Hydrogen Bonded Polymers
Volume Editor: Binder, W.
Vol. 207, 2007

Oligomers · Polymer Composites · Molecular Imprinting
Vol. 206, 2007

Polysaccharides II
Volume Editor: Klemm, D.
Vol. 205, 2006

Neodymium Based Ziegler Catalysts – Fundamental Chemistry
Volume Editor: Nuyken, O.
Vol. 204, 2006

Polymers for Regenerative Medicine
Volume Editor: Werner, C.
Vol. 203, 2006

Peptide Hybrid Polymers
Volume Editors: Klok, H.-A., Schlaad, H.
Vol. 202, 2006

Supramolecular Polymers · Polymeric Betains · Oligomers
Vol. 201, 2006

Ordered Polymeric Nanostructures at Surfaces
Volume Editor: Vancso, G. J., Reiter, G.
Vol. 200, 2006

Emissive Materials · Nanomaterials
Vol. 199, 2006

Surface-Initiated Polymerization II
Volume Editor: Jordan, R.
Vol. 198, 2006

Surface-Initiated Polymerization I
Volume Editor: Jordan, R.
Vol. 197, 2006

Conformation-Dependent Design of Sequences in Copolymers II
Volume Editor: Khokhlov, A. R.
Vol. 196, 2006

Conformation-Dependent Design of Sequences in Copolymers I
Volume Editor: Khokhlov, A. R.
Vol. 195, 2006

Enzyme-Catalyzed Synthesis of Polymers
Volume Editors: Kobayashi, S., Ritter, H., Kaplan, D.
Vol. 194, 2006

Polymer Therapeutics II
Polymers as Drugs, Conjugates and Gene Delivery Systems
Volume Editors: Satchi-Fainaro, R., Duncan, R.
Vol. 193, 2006

Polymer Therapeutics I
Polymers as Drugs, Conjugates and Gene Delivery Systems
Volume Editors: Satchi-Fainaro, R., Duncan, R.
Vol. 192, 2006

Interphases and Mesophases in Polymer Crystallization III
Volume Editor: Allegra, G.
Vol. 191, 2005

Block Copolymers II
Volume Editor: Abetz, V.
Vol. 190, 2005

Block Copolymers I
Volume Editor: Abetz, V.
Vol. 189, 2005

Intrinsic Molecular Mobility and Toughness of Polymers II
Volume Editor: Kausch, H.-H.
Vol. 188, 2005

Oligomers · Polymer Composites · Molecular Imprinting

With contributions by
B. Boutevin · C. Boyer · I. Csetneki · G. David
J. S. Ferguson · G. Filipcsei · B. Gong · S. Li · W. Li
A. R. Sanford · A. Szilágyi · M. Zrínyi

 Springer

The series *Advances in Polymer Science* presents critical reviews of the present and future trends in polymer and biopolymer science including chemistry, physical chemistry, physics and material science. It is adressed to all scientists at universities and in industry who wish to keep abreast of advances in the topics covered.

As a rule, contributions are specially commissioned. The editors and publishers will, however, always be pleased to receive suggestions and supplementary information. Papers are accepted for *Advances in Polymer Science* in English.

In references *Advances in Polymer Science* is abbreviated *Adv Polym Sci* and is cited as a journal.

Springer WWW home page: springer.com
Visit the APS content at springerlink.com

Library of Congress Control Number: 2006934209

ISSN 0065-3195
ISBN 978-3-540-46829-5 Springer Berlin Heidelberg New York
DOI 10.1007/978-3-540-46830-1

This work is subject to copyright. All rights are reserved, whether the whole or part of the material is concerned, specifically the rights of translation, reprinting, reuse of illustrations, recitation, broadcasting, reproduction on microfilm or in any other way, and storage in data banks. Duplication of this publication or parts thereof is permitted only under the provisions of the German Copyright Law of September 9, 1965, in its current version, and permission for use must always be obtained from Springer. Violations are liable for prosecution under the German Copyright Law.

Springer is a part of Springer Science+Business Media

springer.com

© Springer-Verlag Berlin Heidelberg 2007

The use of registered names, trademarks, etc. in this publication does not imply, even in the absence of a specific statement, that such names are exempt from the relevant protective laws and regulations and therefore free for general use.

Cover design: WMXDesign GmbH, Heidelberg
Typesetting and Production: LE-TEX Jelonek, Schmidt & Vöckler GbR, Leipzig

Printed on acid-free paper   02/3100 YL – 5 4 3 2 1 0

Editorial Board

Prof. Akihiro Abe
Department of Industrial Chemistry
Tokyo Institute of Polytechnics
1583 Iiyama, Atsugi-shi 243-02, Japan
aabe@chem.t-kougei.ac.jp

Prof. A.-C. Albertsson
Department of Polymer Technology
The Royal Institute of Technology
10044 Stockholm, Sweden
aila@polymer.kth.se

Prof. Ruth Duncan
Welsh School of Pharmacy
Cardiff University
Redwood Building
King Edward VII Avenue
Cardiff CF 10 3XF, UK
DuncanR@cf.ac.uk

Prof. Karel Dušek
Institute of Macromolecular Chemistry,
Czech
Academy of Sciences of the Czech Republic
Heyrovský Sq. 2
16206 Prague 6, Czech Republic
dusek@imc.cas.cz

Prof. W. H. de Jeu
FOM-Institute AMOLF
Kruislaan 407
1098 SJ Amsterdam, The Netherlands
dejeu@amolf.nl
and Dutch Polymer Institute
Eindhoven University of Technology
PO Box 513
5600 MB Eindhoven, The Netherlands

Prof. Jean-François Joanny
Physicochimie Curie
Institut Curie section recherche
26 rue d'Ulm
75248 Paris cedex 05, France
jean-francois.joanny@curie.fr

Prof. Hans-Henning Kausch
Ecole Polytechnique Fédérale de Lausanne
Science de Base
Station 6
1015 Lausanne, Switzerland
kausch.cully@bluewin.ch

Prof. Shiro Kobayashi
R & D Center for Bio-based Materials
Kyoto Institute of Technology
Matsugasaki, Sakyo-ku
Kyoto 606-8585, Japan
kobayash@kit.ac.jp

Prof. Kwang-Sup Lee
Department of Polymer Science &
Engineering
Hannam University
133 Ojung-Dong
Daejeon 306-791, Korea
kslee@hannam.ac.kr

Prof. L. Leibler
Matière Molle et Chimie
Ecole Supérieure de Physique
et Chimie Industrielles (ESPCI)
10 rue Vauquelin
75231 Paris Cedex 05, France
ludwik.leibler@espci.fr

Prof. Timothy E. Long
Department of Chemistry
and Research Institute
Virginia Tech
2110 Hahn Hall (0344)
Blacksburg, VA 24061, USA
telong@vt.edu

Prof. Ian Manners
School of Chemistry
University of Bristol
Cantock's Close
BS8 1TS Bristol, UK
ian.manners@bristol.ac.uk

Prof. Martin Möller
Deutsches Wollforschungsinstitut
an der RWTH Aachen e.V.
Pauwelsstraße 8
52056 Aachen, Germany
moeller@dwi.rwth-aachen.de

Prof. Oskar Nuyken
Lehrstuhl für Makromolekulare Stoffe
TU München
Lichtenbergstr. 4
85747 Garching, Germany
oskar.nuyken@ch.tum.de

Prof. E. M. Terentjev
Cavendish Laboratory
Madingley Road
Cambridge CB 3 OHE, UK
emt1000@cam.ac.uk

Prof. Brigitte Voit
Institut für Polymerforschung Dresden
Hohe Straße 6
01069 Dresden, Germany
voit@ipfdd.de

Prof. Gerhard Wegner
Max-Planck-Institut
für Polymerforschung
Ackermannweg 10
Postfach 3148
55128 Mainz, Germany
wegner@mpip-mainz.mpg.de

Prof. Ulrich Wiesner
Materials Science & Engineering
Cornell University
329 Bard Hall
Ithaca, NY 14853, USA
ubw1@cornell.edu

Advances in Polymer Science
Also Available Electronically

For all customers who have a standing order to Advances in Polymer Science, we offer the electronic version via SpringerLink free of charge. Please contact your librarian who can receive a password or free access to the full articles by registering at:

springerlink.com

If you do not have a subscription, you can still view the tables of contents of the volumes and the abstract of each article by going to the SpringerLink Homepage, clicking on "Browse by Online Libraries", then "Chemical Sciences", and finally choose Advances in Polymer Science.

You will find information about the

- Editorial Board
- Aims and Scope
- Instructions for Authors
- Sample Contribution

at springer.com using the search function.

Contents

Enforced Folding of Unnatural Oligomers:
Creating Hollow Helices with Nanosized Pores
B. Gong · A. R. Sanford · J. S. Ferguson   1

Telechelic Oligomers and Macromonomers by Radical Techniques
B. Boutevin · G. David · C. Boyer 31

Magnetic Field-Responsive Smart Polymer Composites
G. Filipcsei · I. Csetneki · A. Szilágyi · M. Zrínyi 137

Molecular Imprinting: A Versatile Tool
for Separation, Sensors and Catalysis
W. Li · S. Li 191

Author Index Volumes 201–206 211

Subject Index 215

Enforced Folding of Unnatural Oligomers: Creating Hollow Helices with Nanosized Pores

Bing Gong (✉) · Adam R. Sanford · Joseph S. Ferguson

Department of Chemistry, University at Buffalo, The State University of New York, Buffalo, NY 14260, USA
bgong@chem.buffalo.edu

1	Introduction	2
2	Backbone-Rigidified Folding Oligoamides with Tunable Cavity Sizes	3
2.1	A Novel Three-Center Hydrogen Bond	3
2.2	Crescent Oligoamides	7
2.3	Hollow Helices Based on Rigidified Structural Motifs	9
2.4	Tuning Cavity Size by Adjusting Backbone Curvature	12
2.5	Synthesis: Efficient Coupling Chemistry Based on Acid Chlorides and Aromatic Amines	13
2.6	Folding-Assisted Macrocyclization: Highly Efficient Formation of Shape-Persistent Oligoamide Macrocycles	15
2.7	Highly Specific Receptors for the Guanidinium Ion	17
3	Porous Structures Based on the Enforced Folding of Other Backbones	19
3.1	Enforced Folding of Oligo(m-phenylene ethynylenes)	20
3.2	Folding-Assisted Aromatic Tetraurea Macrocycles	21
3.3	Cavity-Containing Tetrasulfonamide Macrocycles	22
4	Other Foldamer Systems with Rigidified Backbones	23
4.1	Oligoanthranilimides	23
4.2	Folding Oligoamides Consisting of Heterocyclic Residues	24
4.3	Other Backbone-Rigidified Helical Foldamers with Interior Cavities	25
5	Conclusions	26
	References	27

Abstract This article reviews the progress made during the last several years on developing folding helical oligomers consisting of aromatic residues based on a backbone-rigidification strategy. In this approach, rigid, planar, aromatic residues are linked by planar linkers such as amide and urea functionalities. Folding into helical conformations is realized based on the incorporation of localized intramolecular hydrogen bonds that limit the rotational freedom of the backbones, and through the introduction of backbone curvature by linking the aromatic residues in a nonlinear fashion. As a result, the corresponding oligomers are forced to adopt well-defined helical conformations containing interior cavities. Changing the backbone curvature leads to the tuning of the cavity sizes. Such enforced folding based on a "tying up" leads to stable helical backbones that are independent of side chain substitution, which offers a variety of robust helical scaffolds for presenting functional groups. The pore-containing helices combine the feature of secondary and tertiary structures, a feature seen in few other natural or unnatural folding systems. Such an enforced folding approach should provide a simple, predictable strategy for developing a new class of nanoporous structures with predictable dimensions.

Keywords Aromatic oligoamides · Foldamer · Folding · Helix · Nanoporous

Abbreviations

AFM	Atomic force microscopy
CD	Circular dichroism
1D	One-dimensional
2D	Two-dimensional
DCC	Dicyclohexylcarbodiimide
DMF	N,N-Dimethylformamide
DMSO	Dimethyl sulfoxide
DNA	Deoxyribonucleic acid
EDC	1-Ethyl-3-(3-dimethylaminopropyl) carbodiimide
ESI	Electrospray ionization
GPC	Gel-permeation chromatography
HATU	1-[Bis(dimethylamino)methylene]-1H-1,2,3-triazolo[4,5-b]pyridinium-3-oxide hexafluorophosphate
IR	Infrared
MALDI	Matrix-assisted laser desorption/ionization
NMR	Nuclear magnetic resonance
NOE	Nuclear Overhauser enhancement
NOESY	Nuclear Overhauser enhancement (effect) spectroscopy
PE	Phenylene ethynylene
m-PEs	meta-Phenylene ethynylenes
TFA	Trifluoroacetyl
T_m	Melting temperature of DNA
UV	Ultraviolet

1
Introduction

During the last 10 years there has been increasing interest in the design and synthesis of both oligomeric and polymeric systems that fold into well-defined secondary structures [1–10]. Foldamer research is the culmination of efforts beginning in the early twentieth century at the advent of synthetic polymer chemistry and the growth of fields such as molecular biology and supramolecular biology [4]. The amalgamation of these fields has led to the development of unnatural folding systems.

The field of unnatural folding oligomers (aka foldamers) was pioneered by the research groups of Gellman and Seebach, from the studies of secondary structural properties of β-peptides [11–23]. A foldamer, as defined by Moore, is "any oligomer that folds into a conformationally ordered state in solution, the structures of which are stabilized by a collection of noncovalent interactions between nonadjacent monomer units" [4]. More simply, a foldamer can be expressed as an unnatural system that adopts a well-defined secondary structure [2].

Numerous unnatural oligomers that fold into these well-defined secondary structures (foldamers) have been reported since the pioneering work of Gellman and Seebach [11, 19–21]. Recent progress has been made in designing pseudo-biological oligomers (or polymers) that fold predictably. Major systems involve those developed in the laboratories of Iverson [23–26], Lehn [27–30], Moore [31–47], Gong [48–55], and Huc [56–65]. Indeed, since the mid-1990s a veritable explosion in the development of novel folding systems has opened the door to wide-ranging biological and materials science applications. Unnatural foldamers are being developed to answer whether nature has the monopoly on folded structures, to mimic the functions of biomacromolecules, to test if new functions not seen in nature can be developed, and finally, to probe whether diverse chemical challenges can be met [2].

Among reported foldamer systems, those consisting of flat, rigid, aromatic residues constitute a significant percentage [8, 9]. Most of these oligomers fold into various helical conformations. This review summarizes the progress made in recent years on designing and preparing porous structures based on a backbone-rigidification strategy that enforces crescent or helical conformations on aromatic oligoamides and related oligomers. Our own effort along this direction has led to foldamers with enforced, well-defined helical conformations containing cavities of adjustable sizes. Recently, the enforced folding of precursor oligomers was also found to facilitate highly efficient macrocyclization processes, leading to shape-persistent macrocycles with large interior cavities [66].

2
Backbone-Rigidified Folding Oligoamides with Tunable Cavity Sizes

2.1
A Novel Three-Center Hydrogen Bond

We have developed a general strategy for achieving a folded conformation based on the rigidification of the backbones of aromatic oligoamides. The key to the success of this approach is a three-center hydrogen bond consisting of the five- and six-membered hydrogen-bonded rings as shown by the general structure 1. We extensively investigated the stability of this previously unknown three-center hydrogen bond using both theoretical [49, 67, 68] and experimental [48–52, 66] methods.

Amide **1a** and its structural isomers **1b–d** were studied using the Gaussian 98 program, X-ray crystallography, and IR and ^{1}H NMR spectroscopies. The obtained results clearly revealed *positive cooperativity* between the two-center hydrogen-bonding components, i.e., the two two-center components of the three-center hydrogen bond reinforce each other.

The calculated and experimental results, shown in Table 1, closely follow the same trend. The cooperative effects are demonstrated by the N–H stretching frequencies. Relative to **1d**, ab initio calculation indicates that the formation of a two-center H-bond in a five-membered ring in **1c** produces a small blueshift of the N–H stretching frequency, indicating the weakness of this two-center H-bond which alone does not have the strength to produce the expected redshift. IR experiment showed that the N–H stretching fre-

Table 1 Calculated and measured parameters for the N–H bonds of **1a–d** [a]

Compd	ν_{NH} (calcd) [b] (cm^{-1})	ν_{NH} (obsd) (cm^{-1})	δ_{NH} (calcd) [c] (ppm)	δ_{NH} (obsd) [c] (ppm)	r_{NH} (calcd) (Å)	r_{NH} (obsd) (Å)
1a	3423 (−57)	3340 (−104)	10.43 (3.16)	10.61 (3.00)	1.013 (0.005)	0.91±0.02
1b	3437 (−43)	3372 (−72)	9.80 (2.53)	9.61 (2.00)	1.011 (0.003)	0.86±0.02
1c	3485 (+5)	3433 (−11)	8.46 (1.19)	8.50 (0.90)	1.009 (0.001)	0.85±0.02
1d [d]	3480 (0)	3444 (0)	7.27 (0.00)	7.61 (0.00)	1.008 (0.000)	–

[a] Values in parentheses relative to **1d'**. IR and NMR measurements carried out at 1 mM of sample concentration
[b] Frequencies scaled by 0.9613
[c] Values of chemical shifts relative to that of TMS (calculated value = 31.91 ppm)
[d] Calculation based on **1d**; IR and NMR measurements based on **1d'**

quency of **1c** did show a small but clearly discernable redshift of 11 cm^{-1}. In sharp contrast, the cooperative action of the two two-center components in **1a** leads to a much larger redshift in the N–H stretching frequency of **1a** than the two-center bond in **1b** or **1c**. The sizable redshift value clearly indicates nonpairwise effects.

Cooperative effects are also manifested by the ^{1}H NMR signals of the amide protons, as indicated by results from both computational and experimental studies. There is a shift to lower field upon formation of an H-bond. The shift is larger for the two-center H-bond of **1b** than for that of **1c**, and is the largest upon forming the three-center H-bond in **1a**, indicating the energetic superiority of the three-center H-bonding over the two-center H-bonding.

Ab initio calculation indicated that the donor N–H bonds of amides **1a–c** were all elongated upon formation of intramolecular H-bonds relative to **1d**. The formation of a three-center H-bond in **1a** leads to the largest increase of its N–H bond length among amides **1–3**. In contrast, the N–H bond of **1c** shows the smallest increase in its length, indicating that the two-center H-bond in **1c** is the weakest. The measured N–H bond lengths of **1a**, **1b**, and **1c** based on their crystal structures are consistent with the ab initio results.

The enhanced stability of the three-center hydrogen bond was further confirmed by rates of hydrogen–deuterium (H-D) exchange involving the amide protons. As shown in Fig. 1, compared to those of **1b** and **1c**, the amide proton of **1a** exhibits a long half-life of H-D exchange, suggesting a very slow exchange reaction. Such a result implies that the enforced conformation of **1a** is quite robust, undergoing slow interconversion between conformations.

The intramolecular three-center H-bonding interaction also persists in the solid state. As shown in Fig. 2, in the crystal structure of **1a**, the NH group is involved in both a five-membered ring H-bond and a six-membered ring H-bond, leading to a planar molecule reminiscent of a typical three-center H-bond [70]. The six-membered ring H-bond in **1b** is preserved in its solid-state structure. Amide **1b** adopts a nearly flat conformation due to the presence of this two-center intramolecular H-bond. The five-membered ring H-bond in **1c**, which persists in solution as indicated by IR and NMR

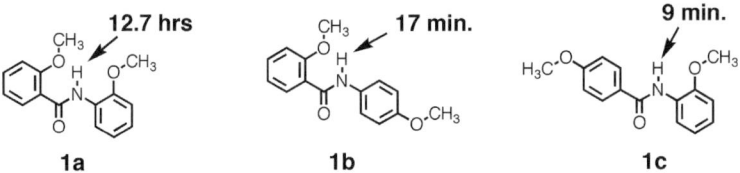

Fig. 1 The half-lives of amide proton–deuterium exchange of amides **1a–c** based on 1D NMR experiments (500 MHz, CDCl$_3$: DMSO-d_7 : D$_2$O = 2 : 19 : 19)

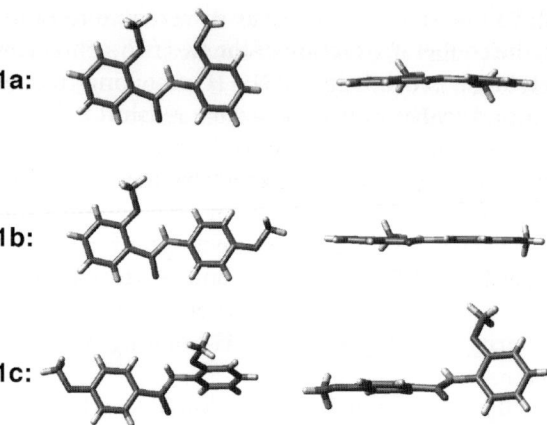

Fig. 2 The crystal structures of amides **1a–c** as viewed from the front (*left*) and side (*right*)

data, is disrupted in the solid state. Instead of forming an intramolecular H-bond, the NH of **1c** is involved in intermolecular H-bonding, indicating the marginal stability of the five-membered ring H-bond observed in solution. Positive cooperativity in the three-center H-bonding in **1a** is clearly indicated by these results; the presence of the six-membered ring component helps the formation of the five-membered ring component which would otherwise be disrupted, as shown by the solid-state structure of **1c**.

The extraordinary stability of the hydrogen-bonded diarylamide structure was further demonstrated by the nuclear Overhauser enhancement spec-

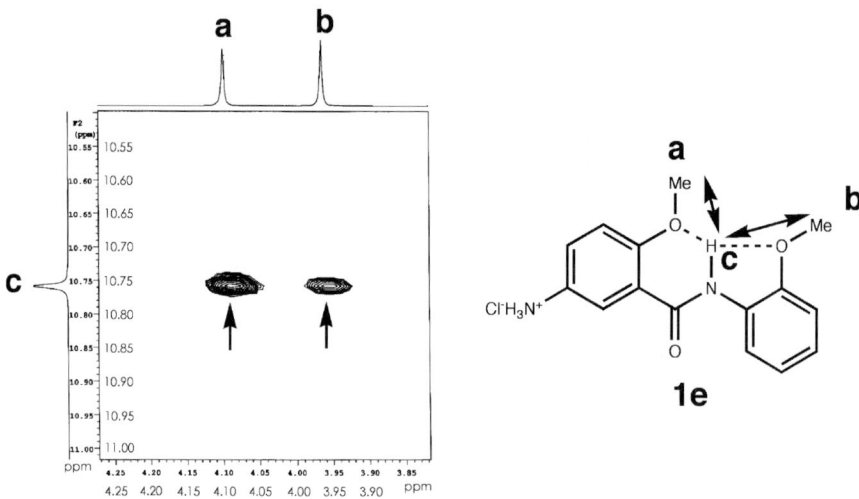

Fig. 3 NOEs observed in the NOESY spectrum of **1e** (6 mM) in 50% DMSO-d_6/50% H_2O (500 MHz, mixing time 0.5 s, 298 K)

troscopy (NOESY) spectrum of the water-soluble amide **1e** recorded in 1 : 1 D_2O and DMSO-d_6 (Fig. 3). The persistence of the three-center hydrogen bond was revealed by two strong NOEs corresponding to contacts between the amide proton and those of the two adjacent methoxy groups. Even in the presence of 50% D_2O, the amide proton of **1e** underwent very slow exchange, which allowed the NOEs involving the amide proton to be recorded. The high stability of the three-center H-bonds in longer oligomers was also demonstrated by the half-lives of amide H-D exchange, ranging from days to too long to be measurable [48, 52].

2.2
Crescent Oligoamides

We designed a series of short aromatic oligoamides with enforced, crescent backbones. Figure 4 shows the crystal structures of a dimer, a trimer, and a tetramer, which are the nitro oligomer intermediates for preparing longer oligomers. The backbones of these short oligomers are almost completely planar due to the presence of the three-center hydrogen bonds. The folded conformations were indicated by the strong amide–side chain NOEs that were detected by NOE difference and NOESY spectra in CDCl$_3$ [48–50, 52]. Each of the amide protons shows the expected NOEs with the protons of the methyl or α- and β-methylene groups of its two adjacent side chains. These amide–side chain NOEs were detected in a variety of solvents with oligomers of different lengths. Their consistent presence has served as a convenient indi-

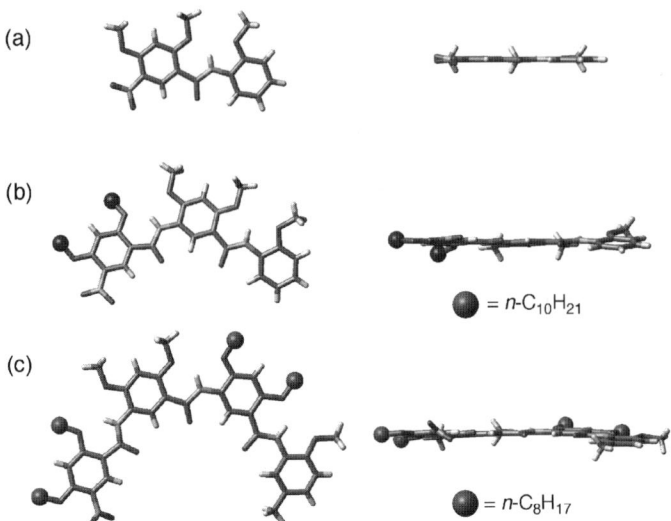

Fig. 4 The crystal structures of **a** a dimer, **b** trimer, and **c** a tetramer, as viewed from front (*left*) and side (*right*)

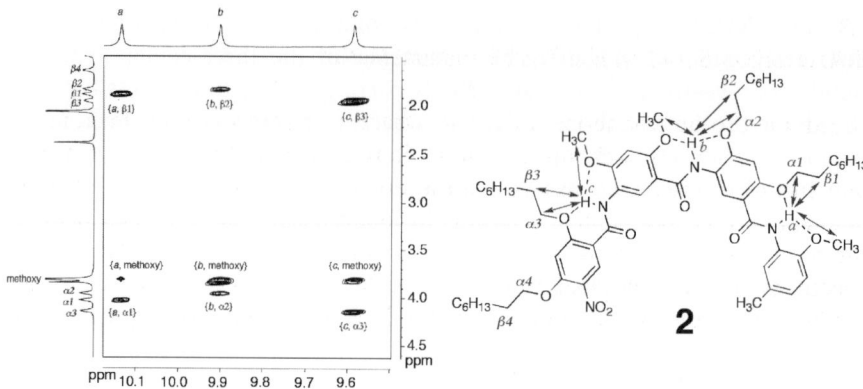

Fig. 5 Expanded plot of the NOESY spectrum of **2** in CDCl$_3$ (50 mM, 800 MHz, 300 K, mixing time: 0.3 s) showing the NOEs between the amide proton and the protons of the side chains

cator of the presence of the three-center H-bonds, and thus for the folding of these crescent-shaped short oligoamides. For example, the NOESY spectrum of tetramer **2** shows three cross peaks for each of the amide protons, corresponding to contacts with protons of the side chains (Fig. 5).

Similar NOESY studies in water (90% H$_2$O, 10% D$_2$O) on water-soluble tetramer **3** showed the same amide–side chain NOEs (Fig. 6), suggesting that these foldamers are stably folded in water. These oligoamides, with rigidified backbones and persistent shape, possess large, amide oxygen-decorated cavities that should serve as hosts for large cations and polar organic molecules.

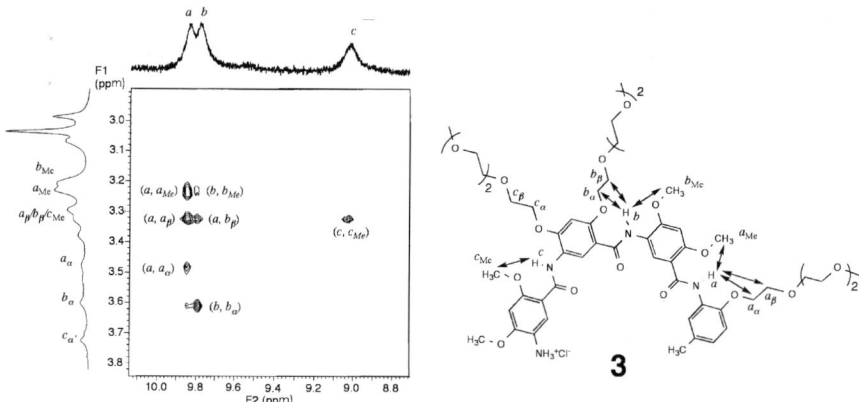

Fig. 6 Partial NOESY spectrum of tetramer **3** in water containing 5% D$_2$O (10 mM, 500 MHz, 297 K, mixing time: 0.5 s). The NOEs between the protons of the amide groups and those of the adjacent side chains are indicated by *arrows*

2.3
Hollow Helices Based on Rigidified Structural Motifs

Based on the backbone-rigidification strategy, oligomers with sufficiently long (≥ 7 subunits) backbones should adopt helical conformations in which one end of the molecule lies above the other. Extensive NOESY studies on the symmetrical nonamer **4** confirmed this expectation [50]. In addition to the presence of numerous well-resolved amide–side chain NOEs, the NOESY spectrum of **4** also revealed an end-to-end NOE cross peak between the end methyl (Me) protons and aromatic proton $b1$ (Fig. 7). Due to the symmetrical

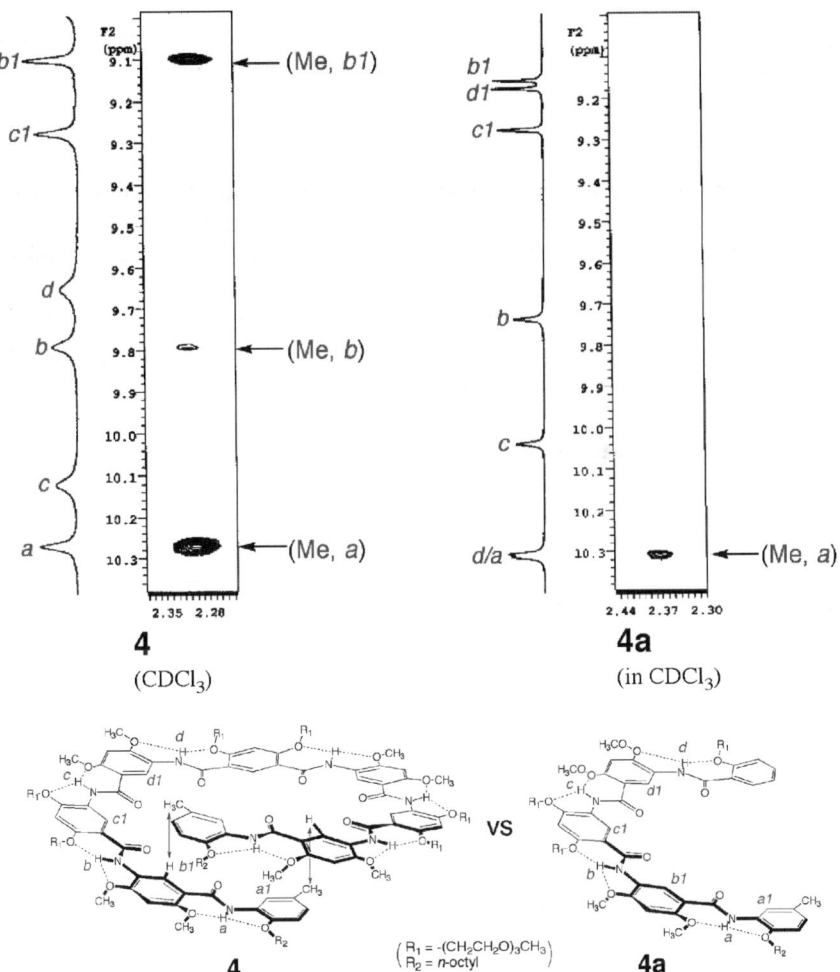

Fig. 7 The helical conformation of nonamer **4** in solution as indicated by the end-to-end NOEs (*arrows*). The same NOEs are not observed for hexamer **4a**

structure of **4**, the observed Me···*b*1 cross peak corresponds to two identical remote NOEs. This NOE can only be explained by a helical conformation adopted by **4**, since the NOESY spectrum of a reference pentamer **4a** (about half of **4**) failed to demonstrate any contact between the corresponding protons. The same Me···*b*1 NOE was also observed for other nonamers that differed with **4** only in their side chains, suggesting that the folding of these oligomers was independent of side chains.

The crystal structure of an analogous symmetrical nonamer **5** (Fig. 8) showed a helical conformation that is fully consistent with the 2D NMR results (Fig. 4b). The molecule contains a large cavity (~ 10 Å across) with disordered solvent molecules (water and DMF). None of the three-center hydrogen bonds was disrupted, although NOEs indicated that some of the five-membered hydrogen-bonded rings were more distorted than the six-membered ones.

Variable-temperature NOESY experiments performed on **4** showed that the Me···*b*1 NOE disappeared much more rapidly than the amide–side chain NOEs (Fig. 9), suggesting that the hydrogen bond rigidified backbone was quite resilient toward heating. Most likely, this molecule "breathes" and extends like a spring but maintains its overall helical conformation.

We recently characterized undecamer **6** using NOESY (Fig. 10), which revealed numerous amide–side chain NOEs that were fully consistent with a three-center hydrogen bond rigidified backbone [52]. Three sets of long-range NOEs (indicated in Fig. 10a), which were not detected in the NOESY spectrum of a hexamer that was about half of **6**, suggest that **6** did indeed adopt a helical conformation.

The rigidity and planar shapes of benzene rings and amide groups, along with the localized three-center H-bonds, suggests that the conformation of

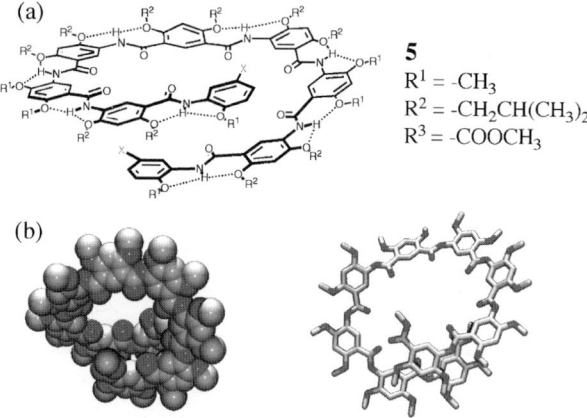

Fig. 8 The crystal structure of nonamer **5** (*left*: space-filling model; *right*: stick model). The side chains are replaced with methyl groups for clarity

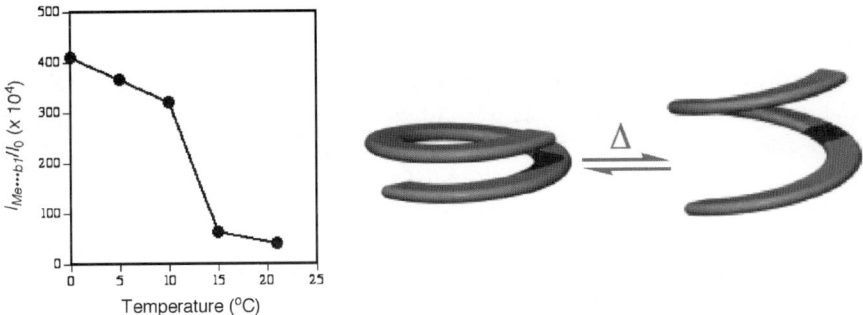

Fig. 9 Change of intensity of the two identical end-to-end (Me···b1) NOEs of nonamer **4** followed by variable-temperature NOESY experiments (2 mM in CDCl$_3$, 500 MHz, mixing time: 140–500 ms), which show an elongation of the folded structure

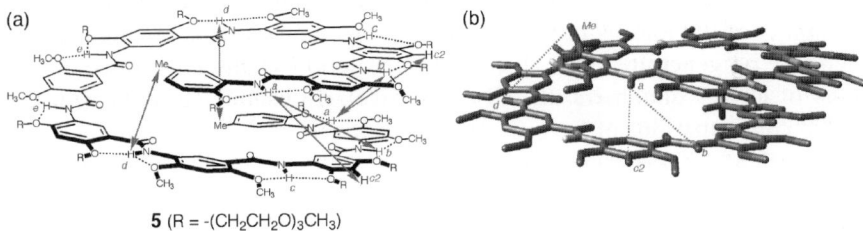

5 (R = -(CH$_2$CH$_2$O)$_3$CH$_3$)

Fig. 10 a Long-range NOEs as revealed by the NOESY spectrum of **6**. **b** Computer-modeled structure of **6** (side replaced with methyl groups) based on the crystal structures of shorter oligomers

the basic structural motif (the diarylamide moiety) is well defined in these foldamers. As a result, the folded conformation of an oligomer can be viewed as the linear combination of all local conformational preferences. Based on the crystal structures of five short oligomers that we reported before [48–50], parameters such as bond lengths, bond angles, dihedral angles, and internuclear distances can be easily extracted. Using these parameters, a simple modeling method that is generally applicable to the prediction of the folded structures of other homologous oligoamides was developed. For example, the average internuclear (C-to-C) distances ($d_{interior}$ and $d_{exterior}$) and amide bond angles (α and β) can serve as convenient structural constraints (Fig. 11).

Thus, restraining $d_{exterior}$ and $d_{interior}$ with the average values shown in Fig. 11 in a modeled structure is in fact a direct reflection of the rigidifying effect of the localized three-center H-bonds, which lead to crescent, tapelike backbones. With the average distances and bond angles, it was found that for *meta*-linked oligomers, it takes about 6.5 residues to complete a full helical turn; for *meta/para*-linked oligomers, a full turn consists of about 19 residues.

Figure 10b shows one such energy-minimized structure corresponding to undecamer **6**. The modeled structure was then examined against the long-

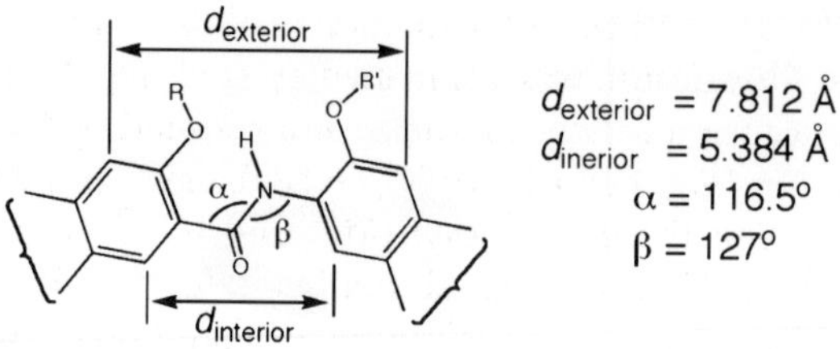

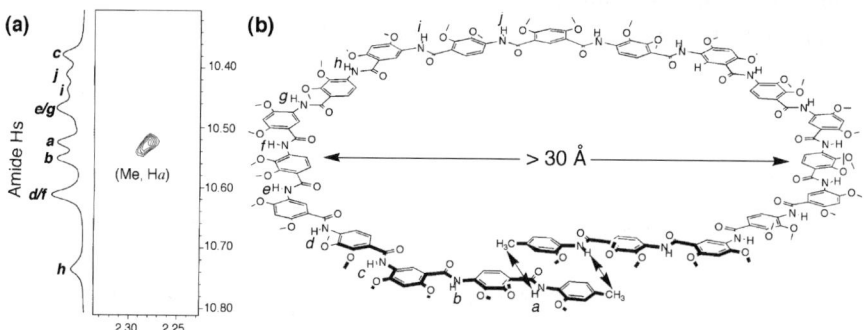

Fig. 11 Average values of distances $d_{exterior}$ and $d_{interior}$, and bond angles α and β determined based on the crystal structures of homologous short oligomers

range NOEs in the NOESY spectrum of **6**. It was found that the modeled structure was not only consistent with the measured NOEs, but also explained the relative intensities of these NOEs. A model of a nonamer corresponding to **4** also matched its NOESY spectrum. These results confirmed the predictability of the folding of these aromatic oligoamides. This method should therefore be generally applicable to the prediction of the folded conformations of longer oligomers. It will be particularly useful for predicting the folded conformations of long oligomers for which characterization methods such NMR and X-ray crystallography fail.

2.4
Tuning Cavity Size by Adjusting Backbone Curvature

By keeping the same three-center hydrogen bonds while incorporating building blocks that place the amide linkages at *para* positions on a benzene ring, the curvature of the backbone of an oligomer should be decreased. Specifically, the backbones of oligomers consisting of alternating building blocks bearing *meta*- and *para*-linked amide residues should be less curved, which

Fig. 12 a End-to-end NOEs as revealed by the NOESY spectrum of **7**. The helical conformation of **7** is consistent with the NOEs (indicated by *purple arrows*). All long side chains are replaced with methyl groups for clarity

is accompanied by an increase of the diameters of the corresponding crescent or helical amide. This design was confirmed by studying 21-mer **7**. The most conclusive evidence for a helical conformation adopted by **7** was provided by its NOESY spectrum (Fig. 12). At −10 °C, an NOE corresponding to two identical contacts between the end methyl (Me) and the first amide (proton *a*) protons was detected. This NOE was absent in a reference decamer which was half of **7**. As mentioned above, modeling shows that it takes about 19 residues to complete a full helical turn, which is confirmed by the NOESY result of **7**. Such a helix has an interior cavity > 30 Å across [50].

By changing the ratio of the *meta*- and *para*-linked building blocks, it should be straightforward to systematically tune the cavities of the corresponding crescent and helical oligomers. Such a strategy is seen in few natural or unnatural systems and is generally applicable to the creation of nanocavities of different sizes.

2.5
Synthesis: Efficient Coupling Chemistry
Based on Acid Chlorides and Aromatic Amines

We have established efficient coupling methods for forming porous aromatic oligoamides in solution [54]. Highly efficient preparations of six different building blocks have been established either in our laboratory (**8a,d,g,h**) or by modifying previously reported procedures (**8b,c,e,f**). Compounds **8a–f** were prepared in hundred-gram quantities and **8f–h** were obtained in multi-gram quantities.

Unsymmetrical oligoamides were synthesized by stepwise (C terminus to N terminus) coupling of the monomer building blocks based on acid chloride chemistry (Scheme 1a). In most cases, except for the first aniline residue, the other residues were incorporated via nitrobenzoic acid chlorides. The resulting nitro-terminated oligomer intermediates were hydrogenated (Pd/C, 4 Pa, in chloroform/ethanol (2 : 1)) into the corresponding amino-terminated analogs which were subjected to the next coupling cycle. The

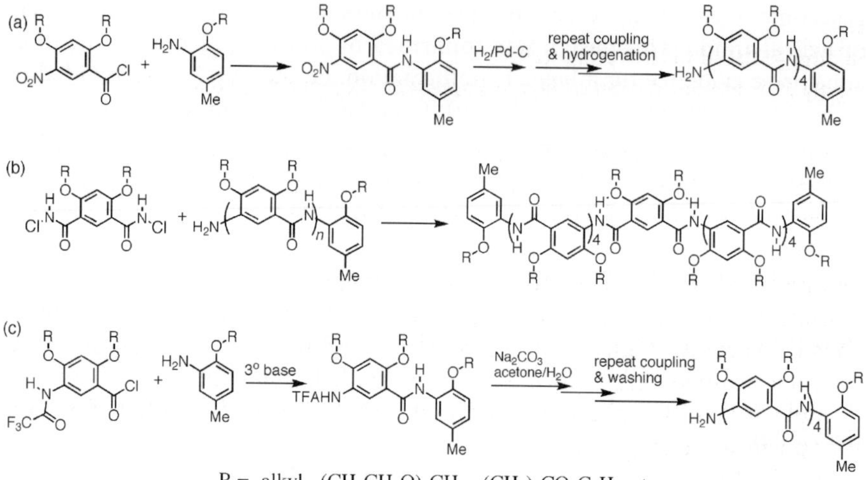

Scheme 1 Synthesis of oligoamides

R = -alkyl, -(CH$_2$CH$_2$O)$_3$CH$_3$, -(CH$_2$)$_3$CO$_2$C$_2$H$_5$, etc.

yield of each coupling step ranged from 75 to 98%. However, as the chain length extends, the coupling efficiency drops and the corresponding purification of the desired products becomes increasingly tedious. To reduce the number of coupling steps, a convergent route involving the coupling of 4,6-disubstituted isophthaloyl chlorides with amino-terminated oligomers was adopted (Scheme 1b). No particular difficulty was encountered in the preparation of the symmetrical oligomers, although the yields of oligomers longer than one helical turn were significantly lower that those of the shorter ones. This is most likely due to the steric hindrance introduced by the folded conformations.

Condensation reagents such as DCC, EDC, and HATU have led to reasonable results in most cases. However, the coupling between monomeric acid chlorides and amino-terminated monomers or oligomers has consistently given the most satisfactory yields.

One disadvantage of the above methods is that the nitro-terminated oligomer intermediates need to be hydrogenated to the amino-terminated oligomers before the next coupling step. As the length of an oligomer increases, this reduction step was found to become increasingly sluggish. Some of the nitro-terminated oligomers were found to have limited solubility in solvents such as CHCl$_3$ and CH$_2$Cl$_2$, which was probably due to strong intermolecular aromatic stacking interactions. To overcome these problems, a trifluoroacetyl (TFA)-protected building block was tested (Scheme 1c). The TFA group can be easily removed under mild basic conditions. It was found that, compared to the corresponding nitro oligomers, the solubilities of TFA-terminated oligomers were significantly improved.

2.6
Folding-Assisted Macrocyclization:
Highly Efficient Formation of Shape-Persistent Oligoamide Macrocycles

In our attempt to prepare AB-type helical polymers based on the backbone-rigidification strategy, diacid chlorides **9** were treated with diamine **10**. Instead of forming polymers, matrix-assisted laser desorption/ionization (MALDI) spectra indicated that macrocycles **11** were consistently formed in very high yields from repeated reactions (Scheme 2) [66].

Under a variety of reaction conditions, such as high or low concentrations of reactants, high or low temperature, and polar or nonpolar solvents, macrocyclic oligomers **11** formed as the overwhelmingly major products. For example, little difference was found in the MALDI spectra of **11a** before and after purification (Fig. 13).

Such highly efficient macrocyclizations are rare since typical one-step, multicomponent macrocyclizations are characterized by poor yields and serious contamination from undesired side products [71]. This discovery has provided a convenient method for preparing large quantities of these macrocycles.

The observed high coupling efficiency can be explained by hydrogen bond enforced preorganization of the uncyclized oligomer precursors. Oligomers

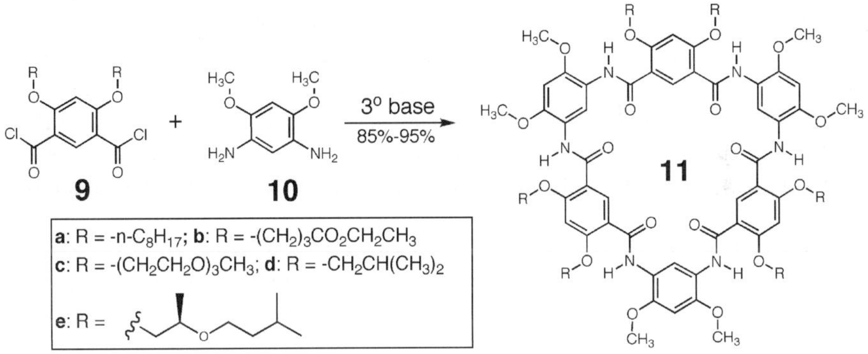

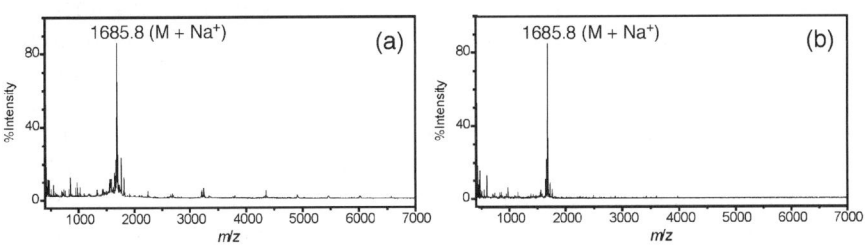

Scheme 2 Synthesis of oligoamide macrocycles

Fig. 13 MALDI-TOF mass spectra of **a** untreated, crude **11a** and **b** purified **11a**

with fewer than six residues could not cyclize easily because of the rigidity of their backbones and the high strain associated with the corresponding macrocycles. For the six-residue oligomer precursor, the folded backbone brought the amino and acid chloride groups into close proximity, resulting in a rapid intramolecular cyclization. Given the irreversible nature of amide bond formation, the predominance of the cyclic hexamers and the scarcity of higher oligomers suggest that the uncyclized six-residue precursors formed more rapidly than other higher oligomers. Such folding-assisted macrocyclization should be generalizable to the preparation of larger macrocycles when *para*-linked residues are included in the design.

It was found that macrocycles **11a** strongly aggregate in various solvents. For example, gel-permeation chromatography (GPC) revealed aggregates with masses up to 200 000 Da (Fig. 14a). The aggregates are quite stable, remaining largely uninterrupted at elevated temperature (Fig. 14b).

The ^{1}H NMR spectra of most of these six-residue macrocycles, particularly those with alkyl side chains, showed severe line broadening. Only at high temperature ($> 60\,^\circ$C) and with mixed solvents containing chloroform-*d* and DMF (or CD_3OD) can the signals be resolved for some of the macrocycles. Atomic force microscopy (AFM) detected very long fibers formed by macrocycle **11a** (Fig. 15a). It was recently found that **11a** acts as a very effective gelator for chloroform, suggesting that the macrocyclic molecules assemble into rodlike objects which then become entangled and trap solvent. This observation indicates that, by incorporating side chains of different properties into these macrocycles, it may be possible to develop a new class of gelators for a wide range of solvents including water.

The most likely mode for the aggregation of macrocycles **11** involves intermolecular aromatic stacking. Face-to-face stacking of the macrocycles can be realized by the relative rotation of adjacent macrocycles, which avoids the

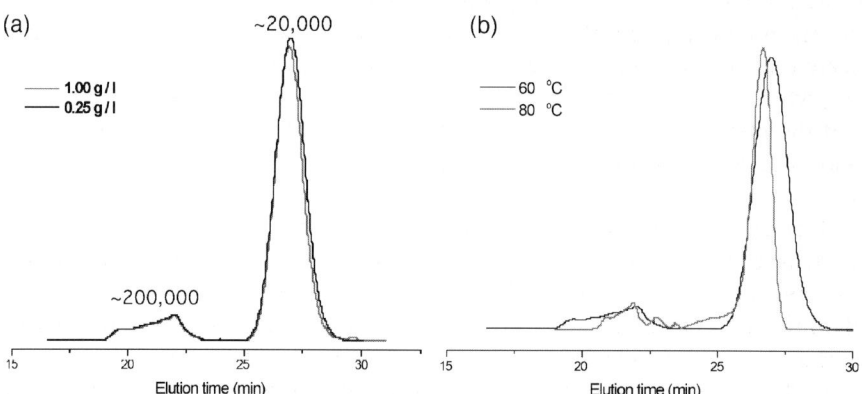

Fig. 14 a GPC trace of **11a** eluted with DMF at 60 °C. **b** Comparison of the GPC traces of **11a** at 60 and 80 °C

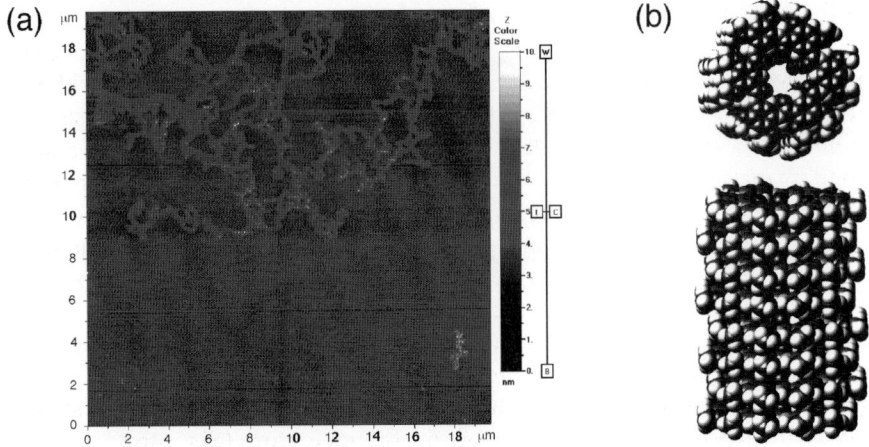

Fig. 15 a AFM shows long fibers formed by **11a**. **b** Proposed stacking of macrocycles **11** (side chain = methyl) into long tubes

stacking of identical residues by placing the electron-rich diamine residue on top of the (relatively) electron-poor diacid residues. Such an arrangement should lead to the alignment of the macrocycles, which results in the formation of the corresponding nanotube with a diameter precisely defined by the sizes of the interior cavities of the constituent macrocycles (Fig. 15b).

2.7
Highly Specific Receptors for the Guanidinium Ion

Simple computer modeling showed that the cavity of macrocycles **11** is almost a tailor-made receptor for binding the guanidinium ion. The six guanidinium hydrogen atoms can form six strong (H $\cdots$ O < 1.9 Å) hydrogen bonds with the six amide oxygen atoms (Fig. 16). In such a rigid complex, it is very likely that the six strong hydrogen bonds act in a positive cooperative fashion, which, in combination with the positive charge of the guanidinium ion and the electrostatically negative interior cavity of the macrocycles, should lead to very strong complexes between the macrocycles and the guanidinium ion [52].

Such an expectation was confirmed first by mass spectrometry. Studies by electrospray ionization (ESI) and MALDI showed that the guanidinium ion was specifically picked up by the six-residue macrocycles. It was found that all of the macrocycles, irrespective of their side chains, complexed the guanidinium ion specifically. With no exception, the stoichiometry observed was 1 : 1. The presence of other cations, such as NH_4^+, NMe_4^+, Li^+, Na^+, K^+, Rb^+, and Cs^+, even in very large (> tenfold) excess, did not interfere with the binding of the guanidinium ion. Figure 17a shows the MALDI spectra of the 1 : 1

Fig. 16 Computer modeling of the 1 : 1 complex of macrocycle **11** (side chain = methyl) and the guanidinium ion

complex between macrocycle **11b** and the guanidinium ion. The presence of other ions did not interfere with the formation of this complex (Fig. 17b).

NOESY studies showed that the guanidinium ion indeed bound in the cavity of the macrocycles (Fig. 18). The observed NOE between the guanidinium NH$_2$ signal and the interior aromatic protons *a* of the diamine residues provided convincing evidence for the binding of the guanidinium ion in the internal cavity of **11b**. Surprisingly, no NOE was detected between the guanidinium protons and the interior aromatic protons *b* of the diacid residues. This suggests the possibility of a nonplanar conformation for macrocycle **11b** when binding the guanidinium ion. In such a conformation, protons *b* are not in the same plane shared by the guanidinium protons, protons *a*, and the amide oxygens, and are probably exposed to solvent molecules.

The discovery of these highly specific receptors has provided the basis for designing specific receptors of substituted guanidinium ions, many of which are biologically relevant [72]. It needs to be pointed out that the noncyclic

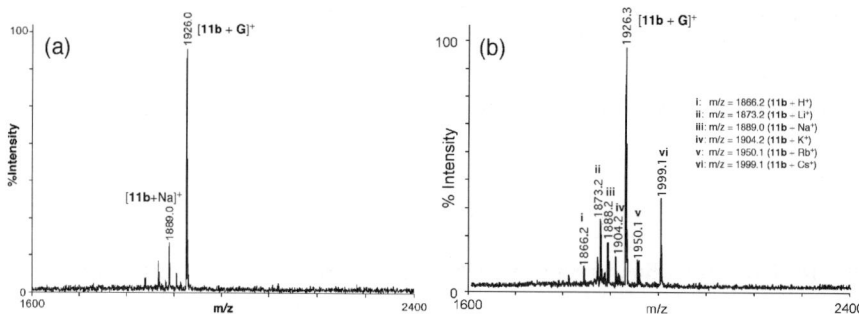

Fig. 17 MALDI-TOF spectra of **a 11b** and guanidinium hydrochloride (G) (1 : 1), and **b 11b** and one equivalent each of guanidine hydrochloride (G), LiCl, NaCl, KCl, RbCl, CsCl, and NH$_4$Cl

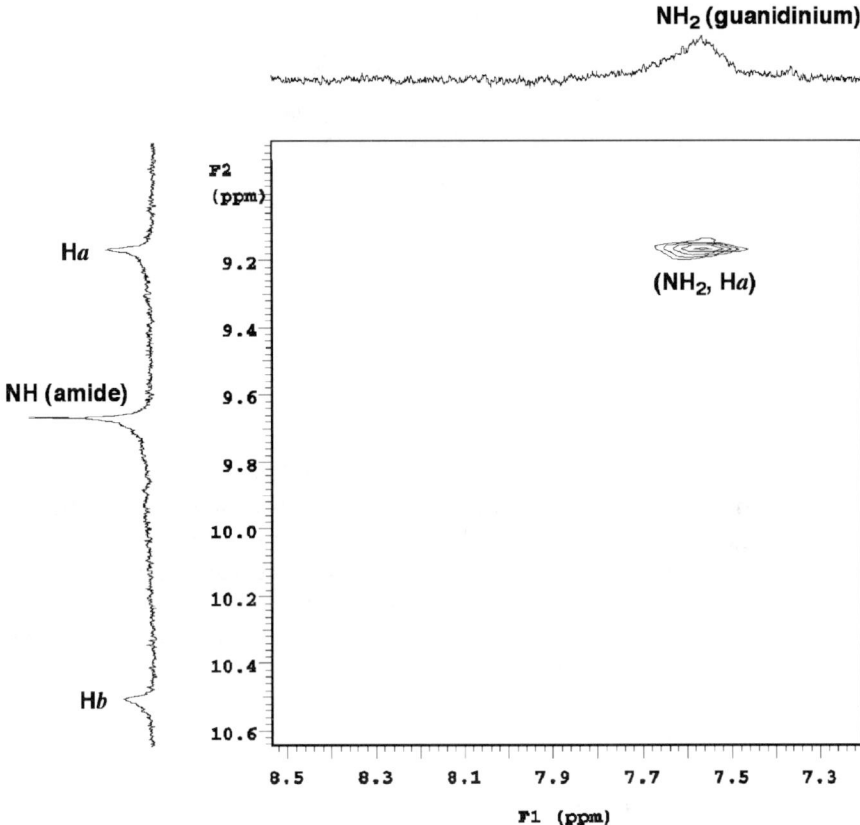

Fig. 18 NOE between the guanidinium (NH$_2$) protons and the aromatic protons a of **11b** as revealed by the NOESY spectrum of the 1:1 mixture of **11b** and guanidinium thiocyanate (500 MHz, 2 mM in acetone-d_6, 283 K, mixing time: 0.4 s)

crescent or helical oligoamides developed by us also possess amide oxygen-rich hydrophilic cavities and should thus act as hosts for large cations and polar molecules. In addition to binding the guanidinium ion and its derivatives, it should be possible to develop receptors for polar molecules such as carbohydrates using our cyclic and folding oligoamides. The tunability of our foldamers should allow tailoring of the cavity size to the corresponding guest.

3
Porous Structures Based on the Enforced Folding of Other Backbones

In addition to the aromatic oligoamides discussed above, we have also designed foldamers by enforcing folded conformations on other oligomers consisting of aromatic residues.

3.1
Enforced Folding of Oligo(*m*-phenylene ethynylenes)

A class of oligo(*m*-phenylene ethynylenes) (oligo(*m*-PEs)) that are forced into folded (crescent or helical) conformations by intramolecular H-bonding were created [51, 53]. The general structure **12** of these *m*-PE foldamers is shown in Fig. 19a. By introducing an intramolecular hydrogen bond, the relatively free internal rotation (0.6 kcal/mol) of the diphenylacetylene unit is limited, which in turn leads to a rigidified backbone. Introducing the same intramolecular hydrogen bond into *m*-PE oligomers of various chain lengths would lead to foldamers with well-defined conformations. Ab initio molecular orbital calculations indicated that **12a** adopted a completely planar conformation that was rigidified by its intramolecular H-bond. The H-bonded conformation **12a** is 5.8 kcal/mol more stable than **12b** (Fig. 19b). Deviation from the planar conformation of **12a** by interrupting the intramolecular hydrogen bond led to a rapid increase in energy. A rotational barrier of 7.19 kcal/mol between conformers **12a** and **12b** was also found (Fig. 19c). The H-bonded conformation **12a** was confirmed by its crystal structure (Fig. 19d). As expected, the intramolecular H-bond leads to a planar conformation that is consistent with results from ab initio calculations.

Oligo(*m*-PEs) **13–18** with two to seven residues were synthesized by stepwise, Pd-catalyzed (Sonogashira) coupling. These oligomers were then char-

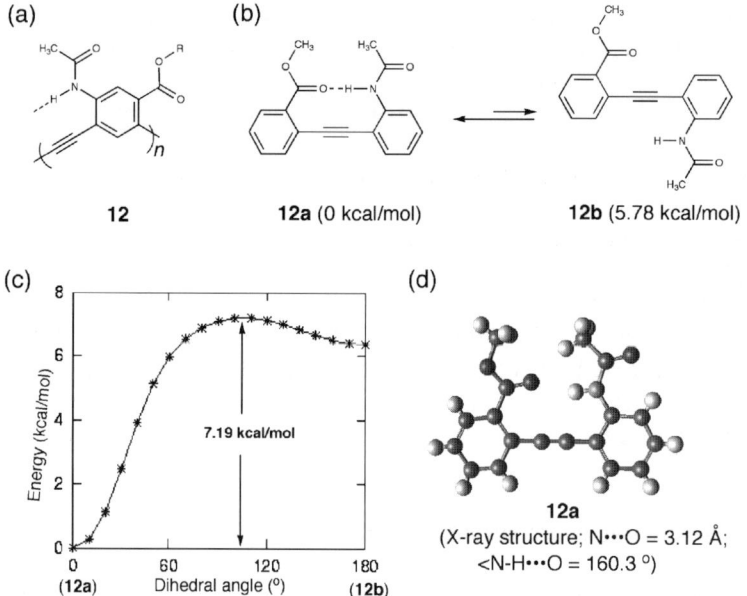

Fig. 19 a Backbone-rigidified *m*-PE foldamers. **b** Energies of **12a** and **12b**. **c** Computed rotational barrier from **12a** to **12b**. **d** The crystal structure of **12a**

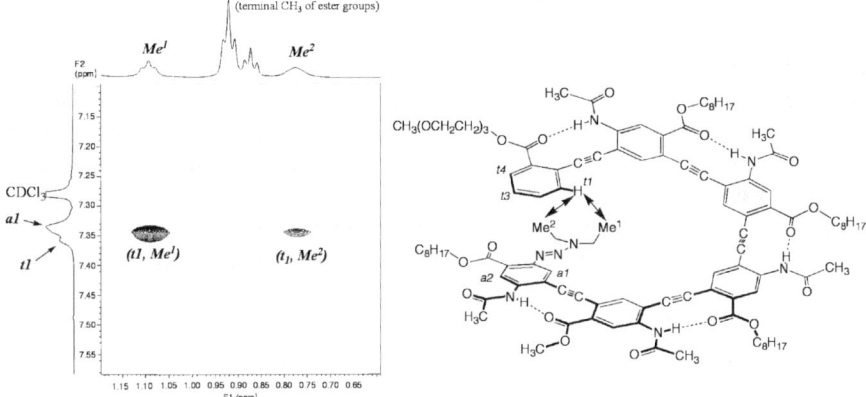

Fig. 20 Partial NOESY spectrum of hexamer **19** (8 mM in CDCl$_3$, 500 MHz, 263 K, mixing time: 0.3 s). The end-to-end contacts between proton $t1$ and the methyl protons Me^1 and Me^2 are indicated in the structure

acterized using 1D and 2D ^{1}H NMR and UV spectroscopy. It was found that pentamer **16**, hexamer **17**, and heptamer **18** adopted well-defined, helical conformations in chloroform, a solvent denaturing m-PE oligomers that fold by solvophobic interaction [31]. Figure 20 shows the expanded NOESY spectrum of hexamer **19**, which reveals NOEs between the remote protons of the two terminal residues. The observed NOEs can only be explained by a helical conformation adopted by **19**. Similar to the folding aromatic oligoamides developed by us, our results on oligo(m-PEs) provided another system of oligomers with predictable folding and well-defined conformations.

3.2
Folding-Assisted Aromatic Tetraurea Macrocycles

Intramolecular hydrogen bonds were also incorporated into aromatic oligo-(ureas) [73]. For example, NMR studies showed that tetramer **20** adopted a crescent conformation. Cyclic tetramer **21** was prepared in high yield

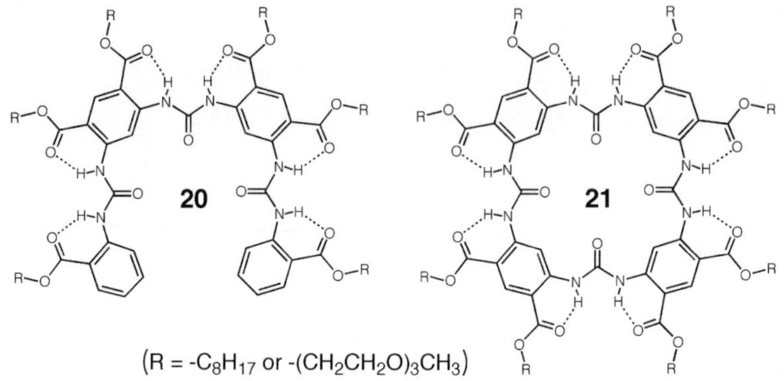

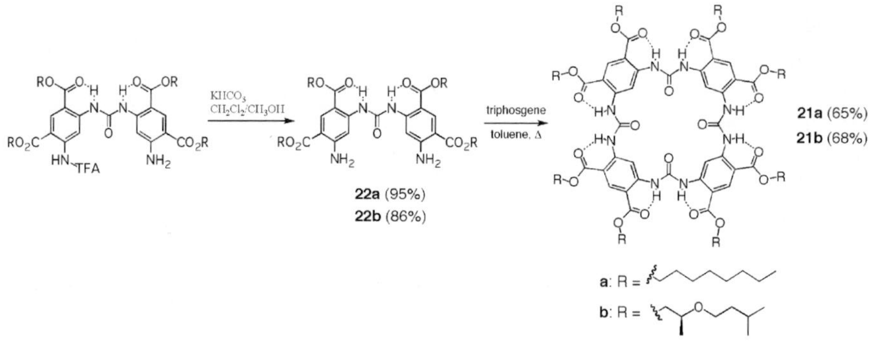

Scheme 3 Synthesis of tetraurea macrocycles

(> 80%) from the corresponding linear precursor (Scheme 3). The cyclic tetraurea macrocycles were more conveniently prepared by a [2 + 2] macrocyclization that involves the dimerization of two dimers [73]. Obviously, the cyclization leading to **21** was assisted by the preorganized backbone of the linear precursors. Binding studies indicated that **21** showed high specificity toward the potassium ion. Even more interesting is the fact that, based on computer modeling, the flat surface area of **21** matches almost exactly the size of the coplanar guanine tetrad found in quadruplex DNA. Preliminary binding studies showed that, indeed, even a neutral **21b** analog (with a chiral ether side chain) dramatically increased the T_m of a target quadruplex DNA ($\Delta T \geq 30\,°C$).

3.3
Cavity-Containing Tetrasulfonamide Macrocycles

In an attempt to develop aromatic oligosulfonamides that fold by backbone-rigidifying hydrogen bonds, we accidentally discovered tetrasulfonamide

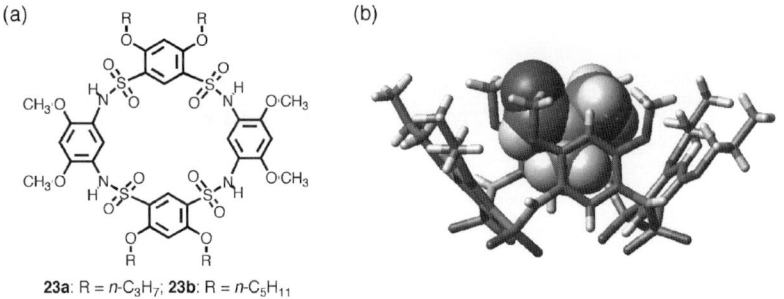

Fig. 21 a Aromatic tetrasulfonamide macrocycles **23a** and **23b** designed and characterized. b The crystal structure of **23a**

macrocycles adopting a cone-shaped conformation (Fig. 21a) that was confirmed by both 2D NMR and X-ray crystallography [74]. The crystal structure of **23a** is shown in Fig. 21b. These tetrasulfonamides are reminiscent of the cone-shaped conformation of calix[4]arenes, with a cavity surrounded by the corresponding aromatic rings. Such a cavity may serve as host for small guest molecules. For example, in the solid state, a DMF molecule was found in the cavity of **23a** (Fig. 21b). Such a well-defined cone conformation may provide a platform for designing cavity-containing molecules of various properties. For example, replacing the numerous phenolic ether side chains should lead to the control of the solubility and even binding behavior of the corresponding macrocyclic molecules. Replacing the benzene residues with larger aromatic residues should lead to the tuning of the cavity sizes.

4
Other Foldamer Systems with Rigidified Backbones

4.1
Oligoanthranilimides

In 1997, Hamilton et al. reported that anthranilimide derivatives could be used as monomer units for helically folding oligomers in the solid state (Fig. 22) [75]. The design is based on (1) the ability of anthranilimide derivatives to form intramolecular hydrogen bonds between adjacent amide-NH and -CO groups, and (2) the backbone was curved by introducing two other subunits based on pyridine-2,6-carboxylic acid and 4,6-dimethoxy-1,3-diaminobenzene. Indeed, in the solid state, a nonamer of this series crystallized into two polymorphs, corresponding to two different two-turn helical structures. In one polymorph, the structure contains two turns of the same helical sense. In the other polymorph, the structure contains one turn of

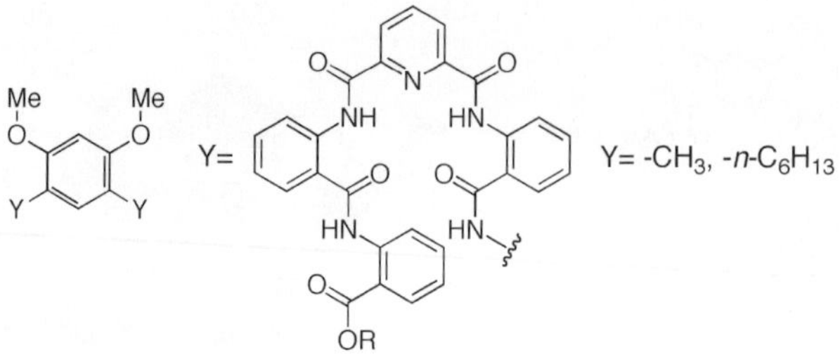

Fig. 22 Nine-residue oligoanthranilimides that fold into helical conformations

a right- and one turn of a left-handed helix linked through the spacer based on the 4,6-dimethoxy-1,3-diaminobenzene subunits.

4.2
Folding Oligoamides Consisting of Heterocyclic Residues

Lehn reported [76] oligomers consisting of 2'-pyridyl-2-pyridinecarboxamide units (Fig. 23). NMR studies indicated that at low concentration, a single species that was consistent with a single-helix structure existed. At increased concentrations, a new species appeared that was attributed to a double helix consisting of two intertwined monomeric strands. The double-stranded helices are stabilized by extensive intermolecular aromatic stacking interactions. Both single and double helices were confirmed by the X-ray structures of less soluble analogs. More recently, based on X-ray diffraction and NMR studies, Huc and coworkers showed that short, oligo(pyridine dicarboxamide) strands bearing benzyloxy and hydroxyl functionalities adopt robust single helical conformations in the solid state. Oligomers bearing hydroxylate groups, which render water solubility, were found by ^{1}H NMR to be folded in polar solvents including water [77]. These oligomers fold via strong intramolecular hydrogen-bonding interactions along the backbone between amide protons and pyridine nitrogens. The resulting helix was reported to be

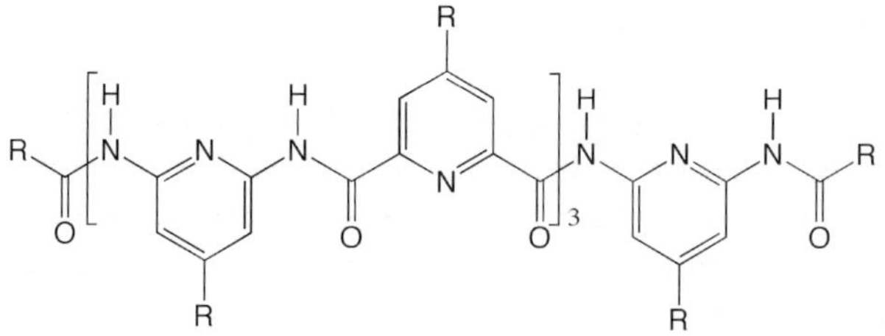

Fig. 23 Oligo(pyridine dicarboxamides). The molecules are forced to adopt helical conformations by intramolecular hydrogen-bonding interactions between amide protons and pyridine nitrogens

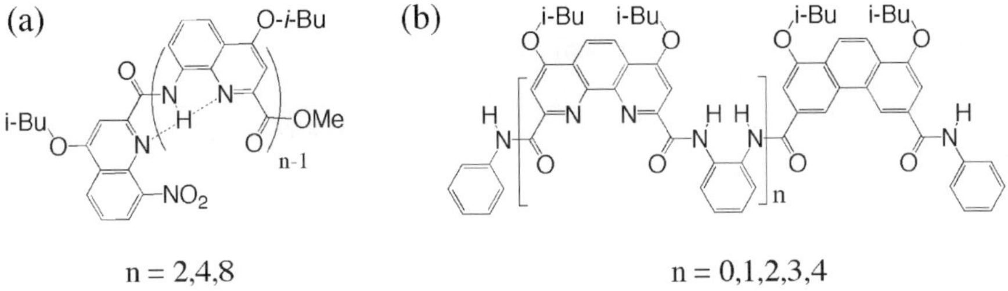

Fig. 24 Backbone-rigidified folding oligoamides described by **a** Huc [65] and **b** Chen [78]

very stable and contained a polar interior channel that showed an affinity to solvent molecules such as water and methanol.

Huc et al. reported aromatic oligoamides based on quinoline residues (Fig. 24a) [65]. It was found that the corresponding oligomers folded into stable helical conformations based on three-center intramolecular hydrogen-bonding interactions involving the amide protons and their adjacent pyridine N atoms. This class of oligomers was found to be stably folded in both nonpolar and polar solvents. Their folded conformations were also revealed by their crystal structures.

Another related system of aromatic oligoamides that fold based on partial backbone rigidification was reported by Chen et al. [78]. This class of foldamers is based on building blocks derived from 1,10-phenanthroline diacid and o-phenylenediamine (Fig. 24b). In spite of the hydrogen-bonded backbones of these oligomers, it was found that methanol promoted the folding of these molecules more than chloroform and methylene chloride, suggesting the role of solvophobic-driven folding in this system.

A partially rigid foldamer system was recently reported by Parquette et al. [79]. The oligomers were derived from alternating sequences of pyridine-2,6-dicarboxamides and *meta*-(phenylazo)azobenzenes. The helical conformations were revealed by crystal structures. In solution, the folded conformations and their corresponding dynamics were probed by NMR.

4.3
Other Backbone-Rigidified Helical Foldamers with Interior Cavities

Except for the folding aromatic oligoamides developed by us, few other helical foldamers reported so far contain any meaningful interior cavities. Unnatural foldamers with interior cavities include those described by Lehn [29] and Moore [31]. A few examples of porous helical oligoamides have started to appear in recent years.

Li et al. recently reported the synthesis and characterization of a series of hydrogen bond-driven hydrazide foldamers and their recognition for alkyl saccharides in chloroform (Fig. 25a) [80]. The rigidity and planarity of this

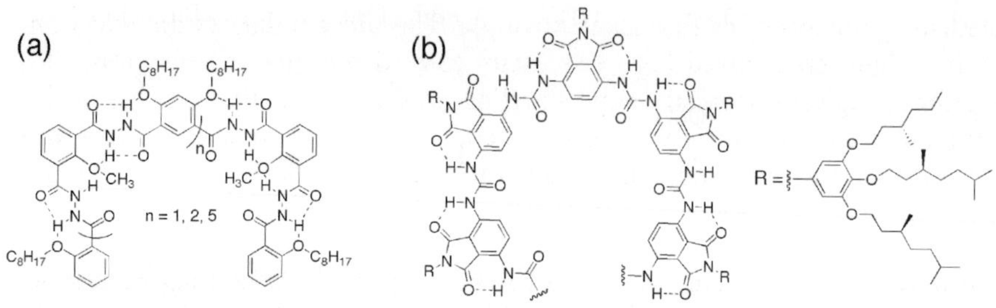

Fig. 25 a The oligohydrazide foldamers developed by Li et al. [80]. **b** The helical polyurea reported by Meijer et al. [81]

system were confirmed by X-ray analysis and ^{1}H NMR of shorter oligomers. Folding and helical conformations of the longer oligomers was determined by 1D and 2D ^{1}H NMR and IR spectroscopy. Molecular mechanics calculations revealed that these oligomers possess a rigid cavity with a diameter size estimated to be 10 Å; half of the carbonyl groups in the folded conformation orient inward toward the inside of the cavity. This feature allowed the system to complex alkylated mono- and disaccharides in chloroform, as evidenced by ^{1}H NMR and circular dichroism (CD) experiments. The association constants of the saccharide complexes were determined using ^{1}H NMR and fluorescence titration methods.

Meijer et al. described oligo- and polyureas that fold into a helical conformation containing a large hydrophilic cavity (Fig. 25b) [81]. Folding in this case is enforced by the intramolecular hydrogen bonds formed between the urea hydrogens and their adjacent imide carbonyl oxygens. CD study showed that this polymer demonstrated a strong Cotton effect in THF, which was explained by transfer of chirality from the UV-silent chiral side chains to the backbone. Interestingly, no Cotton effect was observed for these molecules in chloroform, in which the intramolecular hydrogen bonds should still persist.

5
Conclusions

During the last 5 years, we have fully established the rules for folding a class of porous oligoamides with enforced, curved backbones. The reliability of the three-center H-bond, which plays the critical role in enforcing the folded conformation of these molecules, is demonstrated by all the oligomers characterized. Folding was independent of side chains and solvents. Large hydrophilic internal cavities are obtained and the cavity size can be tuned while maintaining the crescent or helical conformations. Folding of oligomer precursors was found to assist a highly efficient, one-step, multicomponent macrocyclization which leads to large quantities of a new class of shape-persistent macrocycles that stack strongly. The *meta*-linked cyclic and non-cyclic oligomers showed high specificity toward the guanidinium ion. This backbone-rigidification strategy has reliably led to well-defined, stably folded conformations in a variety of solvents and in the solid state. Extending this strategy to other oligomers has also led to stable structures, as shown by our own reports and those of others.

Acknowledgements The authors thank the current and previous Gong group members for their contributions to the research discussed in this article. Work from the authors' laboratory was supported by the National Science Foundation (CHE-0314577), the National Institutes of Health (R01GM63223), and the Office of Naval Research (N000140210519).

References

1. Seebach D, Matthews JL (1997) Chem Commun 2015
2. Gellman SH (1998) Acc Chem Res 31:173
3. Stigers KD, Soth MJ, Nowick JS (1999) Curr Opin Chem Biol 4:714
4. Hill DJ, Mio MJ, Prince RB, Hughes TS, Moore JS (2001) Chem Rev 101:3893
5. Gong B (2001) Chem Eur J 7:4336
6. Cubberley MS, Iverson BL (2001) Curr Opin Chem Biol 5:650
7. Cheng RP, Gellman SH, DeGrado WF (2001) Chem Rev 101:3219
8. Sanford AR, Gong B (2003) Curr Org Chem 7:1649
9. Huc I (2004) Eur J Org Chem 17
10. Sanford AR, Yamato K, Yuan LH, Han YH, Gong B (2004) Eur J Biochem 271:1416
11. Appella DH, Christianson LA, Karle IL, Powell DR, Gellman SH (1996) J Am Chem Soc 118:13071
12. Apella DH, Christianson LA, Klein DA, Powell DR, Huang X, Barchi JJ, Gellman SH (1997) Nature 387:381
13. Applequist J, Bode KA, Appella DH, Christianson LA, Gellman SH (1998) J Am Chem Soc 120:4891
14. Appella DH, Christianson LA, Karle IL, Powell DR, Gellman SH (1999) J Am Chem Soc 121:6206
15. Apella DH, Christianson LA, Klein DA, Richards MR, Powell DR, Gellman SH (1999) J Am Chem Soc 121:7574
16. Appella DH, Barchi JJ, Durell SR, Gellman SH (1999) J Am Chem Soc 121:2309
17. Wang X, Espinosa JF, Gellman SH (2000) J Am Chem Soc 122:4821
18. Barchi JJ, Huang XL, Appella DH, Christianson LA, Durell AR, Gellman SH (2000) J Am Chem Soc 122:2711
19. Seebach D, Overhand M, Kuhnle FNM, Martinoni B, Oberer L, Hommel U, Widmer H (1996) Helv Chim Acta 79:913
20. Seebach D, Ciceri PE, Overhand M, Jaun B, Rigo D (1996) Helv Chim Acta 79:2043
21. Seebach D, Abele S, Gademann K, Guichard G, Hintermann T, Juan B, Mathews JL, Schreiber JV, Oberer L, Hommel U, Widmer H (1998) Helv Chim Acta 81:932
22. Seebach D, Brenner M, Rueping M, Juan B (2002) Chem Eur J 8:573
23. Lokey RS, Iverson BL (1995) Nature 375:303
24. Nguyen JQ, Iverson BL (1999) J Am Chem Soc 121:2639
25. Zych AJ, Iverson BL (2000) J Am Chem Soc 122:8898

26. Gabriel GJ, Iverson BL (2002) J Am Chem Soc 124:15174
27. Hanan GS, Lehn JM, Kyritsakas N, Fischer J (1995) J Chem Soc Chem Commun 765
28. Ohkita M, Lehn JM, Baum G, Fenske D (1999) Chem Eur J 5:3471
29. Cuccia LA, Lehn JM, Homo JC, Schmutz M (2000) Angew Chem Int Ed 39:233
30. Cuccia LA, Ruiz E, Lehn JM, Homo JC, Schmutz M (2002) Chem Eur J 8:3448
31. Nelson JC, Saven JG, Moore JS, Wolynes PG (1997) Science 277:1793
32. Prince RB, Okada T, Moore JS (1999) Angew Chem Int Ed 38:233
33. Gin MS, Yokozawa T, Prince RB, Moore JS (1999) J Am Chem Soc 121:2643
34. Prince RB, Saven JG, Wolynes PG, Moore JS (1999) J Am Chem Soc 121:3114
35. Prest PJ, Prince RB, Moore JS (1999) J Am Chem Soc 121:5933
36. Prince RB, Barnes SA, Moore JS (2000) J Am Chem Soc 122:2758
37. Yang WY, Prince RB, Sabelko J, Moore JS, Gruebele M (2000) J Am Chem Soc 122:3248
38. Mio MJ, Prince RB, Moore JS, Kuebel C, Martin DC (2000) J Am Chem Soc 122:6134
39. Lahiri S, Thompson JL, Moore JS (2000) J Am Chem Soc 122:11315
40. Tanatani A, Mio MJ, Moore JS (2001) J Am Chem Soc 123:1792
41. Brunsveld L, Meijer EW, Prince RB, Moore JS (2001) J Am Chem Soc 123:7978
42. Oh K, Jeong KS, Moore JS (2001) Nature 414:889
43. Tanatani A, Hughes TS, Moore JS (2001) Angew Chem Int Ed 41:325
44. Hill DJ, Moore JS (2002) Proc Natl Acad Sci USA 99:5053
45. Zhao DH, Moore JS (2002) J Am Chem Soc 124:9996
46. Matsuda K, Stone MT, Moore JS (2002) J Am Chem Soc 124:11836
47. Heemstra JM, Moore JS (2004) J Am Chem Soc 126:1648
48. Zhu J, Parra RD, Zeng H, Skrzypczak-Jankun E, Zeng CZ, Gong B (2000) J Am Chem Soc 122:4219
49. Parra RD, Zheng H, Zhu J, Zheng C, Zeng XC, Gong B (2001) Chem Eur J 7:4352
50. Gong B, Zeng H, Zhu Y, Yuan L, Han Y, Cheng S, Furukawa M, Parra RD, Kovalevsky AY, Mills J, Skrzypczak-Jankun E, Martinovic S, Smith RD, Zheng C, Szyperski T, Zeng XC (2002) Proc Natl Acad Sci USA 99:11583
51. Yang XW, Brown AL, Furukawa M, Li SJ, Gardinier WE, Bukowski EJ, Bright FV, Zheng C, Zeng XC, Gong B (2003) Chem Commun 56
52. Yuan LH, Zeng HQ, Yamato K, Sanford AR, Feng W, Atreya HS, Sukumaran DK, Szyperski T, Gong B (2004) J Am Chem Soc 126:16528
53. Yang XW, Yuan LH, Yamamoto K, Brown AL, Feng W, Furukawa M, Zeng XC, Gong B (2004) J Am Chem Soc 126:3148
54. Yuan LH, Sanford AR, Feng W, Zhang AM, Zhu J, Zeng HQ, Yamato K, Li MF, Ferguson JS, Gong B (2005) J Org Chem 70:10660
55. Sanford AR, Yuan LH, Feng W, Yamato K, Flowers RA, Gong B (2005) Chem Commun 4720
56. Dolain C, Léger JM, Delsuc N, Gornitzka H, Huc I (2005) Proc Natl Acad Sci USA 102:16146
57. Dolain C, Grélard A, Laguerre M, Jiang H, Maurizot V, Huc I (2005) Chem Eur J 11:6135
58. Dolain C, Jiang H, Léger JM, Guionneau P, Huc I (2005) J Am Chem Soc 127:12943
59. Jiang H, Maurizot V, Huc I (2004) Tetrahedron 60:10029
60. Jiang H, Léger JM, Guionneau P, Huc I (2004) Org Lett 6:2985
61. Maurizot V, Dolain C, Leydet Y, Léger JM, Guionneau P, Huc I (2004) J Am Chem Soc 126:10049
62. Maurizot V, Linti G, Huc I (2004) Chem Commun 8:924
63. Jiang H, Dolain C, Léger JM, Gornitzka H, Huc I (2004) J Am Chem Soc 126:1034
64. Jiang H, Léger JM, Dolain C, Guionneau P, Huc I (2003) Tetrahedron 59:8365

65. Jiang H, Léger JM, Huc I (2003) J Am Chem Soc 125:3448
66. Yuan LH, Feng W, Yamato K, Sanford AR, Xu DG, Guo H, Gong B (2004) J Am Chem Soc 126:11120
67. Parra RD, Furukawa M, Gong B, Zeng XC (2001) J Chem Phys 115:6030
68. Parra RD, Gong B, Zeng XC (2001) J Chem Phys 115:6036
69. Yang XW, Martinovic S, Smith RD, Gong B (2003) J Am Chem Soc 125:9932
70. Parthasarathy R (1969) Acta Crystallogr B 25:509
71. Krakowiak KE, Izatt RM, Bradshaw JS (2001) J Heterocycl Chem 38:1239
72. Schug KA, Lindner W (2005) Chem Rev 105:67
73. Zhang AM, Han YH, Yamato K, Zeng XC, Gong B (2006) Org Lett 8:803
74. He L, An Y, Yuan LH, Yamato K, Feng W, Gerlitz O, Zheng C, Gong B (2005) Chem Commun 3788
75. Hamilton AD (1997) J Am Chem Soc 119:10587
76. Volker B, Huc I, Khoury RG, Krische MJ, Lehn JM (2000) Nature 407:720
77. Dolain C, Zhan C, Léger JM, Daniels L, Huc I (2005) J Am Chem Soc 127:2400
78. Hu ZQ, Hu HY, Chen CF (2006) J Org Chem 71:1131
79. Chenyang T, Gallucci JC, Parquette JR (2006) J Am Chem Soc 128:1762
80. Hou JL, Shao XB, Chen GJ, Zhou YX, Jiang XK, Li ZT (2004) J Am Chem Soc 126:12386
81. van Gorp JJ, Vekemans JAJM, Meijer EW (2004) Chem Commun 60

Editor: Kwang-Sup Lee

Telechelic Oligomers and Macromonomers by Radical Techniques

B. Boutevin (✉) · G. David (✉) · C. Boyer (✉)

Laboratoire de Chimie Macromoléculaire, UMR/CNRS 5076, Ecole Nationale Supérieure de Chimie de Montpellier, 8 rue de l'école normale, 34296 Montpellier, Cedex 5, France
boutevin@cit.enscm.fr, ghislain.david@enscm.fr, cyrille.boyer@enscm.fr

1	Introduction	33
2	**Synthesis of Telechelic Oligomers by Radical Techniques**	35
2.1	Telechelic Oligomers Obtained by Telomerization	36
2.1.1	Radical Addition	36
2.1.2	Nucleophilic Addition	38
2.1.3	Chain-End Chemical Modification	39
2.2	Telechelic Oligomers Obtained by Dead-End Polymerization	41
2.2.1	Styrene	42
2.2.2	Acrylates	44
2.2.3	Fluoro-type Monomers	46
2.2.4	Other Monomers	47
2.3	Telechelic Oligomers Obtained by Addition–Fragmentation	47
2.3.1	Use of Chain Transfer Agents in Addition–Fragmentation	47
2.3.2	Catalytic Chain Transfer	52
2.4	Telechelic Oligomers Obtained by Other Conventional Radical Polymerizations	53
2.4.1	Use of Initer/Iniferter Systems	53
2.4.2	Oxidative Cleavage	57
2.5	Telechelic Oligomers Obtained by Atom Transfer Radical Polymerzation	58
2.5.1	Synthesis of Telechelic Oligomer Precursors	60
2.5.2	Synthesis of Telechelic Oligomers	61
2.6	Telechelic Oligomers Obtained by Reversible Addition–Fragmentation Chain Transfer	72
2.6.1	Use of a Trithioester Transfer Agent	74
2.6.2	Thioester Modification	76
2.7	Telechelic Oligomers Obtained by Nitroxide-Mediated Polymerization	79
2.7.1	Synthesis of Precursors of Telechelic Oligomers	80
2.7.2	Synthesis of Telechelic Oligomers	84
2.8	Telechelic Oligomers Obtained by Iodine Transfer Polymerization	86
2.8.1	Direct Chemical Change	87
2.8.2	Functionalization by Radical Addition	89
2.8.3	Radical Coupling	89
3	**Synthesis of Macromonomers by Radical Techniques**	90
3.1	New Macromolecular Designs of Macromonomers	91
3.1.1	Acrylic and Styrenic Double Bonds	91
3.1.2	Other Reactive Double Bonds	95
3.1.3	Macromonomers with Polycondensable Groups	96

3.2   Macromonomers Obtained by Telomerization   96
3.2.1 Macromonomers with a Polymerizable Double Bond   98
3.2.2 Macromonomers with Polycondensable Groups   104
3.3   Macromonomers Obtained
       by Addition–Fragmentation and Catalytic Chain Transfer   105
3.3.1 Addition–Fragmentation Process   105
3.3.2 Catalytic Chain Transfer Process   106
3.4   Macromonomers Obtained by Atom Transfer Radical Polymerization . . .   110
3.4.1 Synthesis of Macromonomers with a Polymerizable Double Bond   110
3.4.2 Synthesis of Macromonomers with Polycondensable Groups   115
3.5   Macromonomers Obtained by Nitroxide-Mediated Polymerization   118
3.5.1 Modification of the ω Position   118
3.5.2 Modification of the α Position   119
3.6   Macromonomers Obtained by Other Techniques   121

4   Conclusion   123

References   124

Abstract This review summarizes nearly 400 references (since 1990) intended to highlight directions on the synthesis of telechelic oligomers and macromonomers by radical techniques. This review first takes into account the recent developments in further conventional radical polymerizations, such as dead-end polymerization and also telomerization reactions. Among all the conventional radical polymerizations, addition–fragmentation transfer (AFT) polymerization realized a real breakthrough for the synthesis of telechelic oligomers and especially for macromonomers by coupling AFT with catalytic chain transfer. Then, surveys concerning telechelic oligomers and macromonomers prepared by living radical polymerizations are mentioned. Atom transfer radical polymerization, nitroxide-mediated polymerization, reversible addition–fragmentation chain transfer polymerization and also iodine transfer polymerization allow for accurate control of chain-end functionality, either a functional group or a double bond. Novel reactions like radical coupling of oligomers previously obtained by living radical polymerizations are enhanced.

Keywords Telechelic oligomers · Macromonomers ·
Conventional Radical Polymerizations · Controlled/Living Radical Polymerization ·
Chemical Modification

Abbreviations
ACVA   4,4′-Azobis(4-cyanovaleric acid)
AFT   Addition–fragmentation transfer polymerization
AIBN α,α'-Azobis(isobutyronitrile)
ATRC   Atom transfer radical coupling
ATRP   Atom transfer radical polymerization
BHEBT S,S'-Bis(2-hydroxyethyl-2′-butyrate)trithiocarbonate
BMA   Benzyl methacrylate
n-BA n-Butyl acrylate
CCT   Catalytic chain transfer
Co(tpp)   5,10,15,20-Tetraphenyl-21H,23H-porphine cobalt(II)

CRP	Controlled radical polymerization
CTA	Chain transfer agent
DABCO	1,4-Diazabicyclo[2,2,2]octane
DEP	Dead-end polymerization
DMF	Dimethylformamide
DMSO	Dimethyl sulfoxide
dNbipy	4,4'-Di-(5-nonyl)-2,2'bipyridine
ETPEP	Ethyl-2-[1-((2-tetrahydrofuranyl)peroxy)ethyl]propenoate
FATRIFE	2,2,2-Trifuoroethyl α-fluoroacrylate
GPC	Gel permeation chromatography
HEMA	Hydroxyethyl methacrylate
HMTETA	1,1,4,7,10,10-Hexamethyltriethylenetetraamine
IEM	Isocyanoethyl methacrylate
ITP	Iodine transfer polymerization
LC	Liquid chromatography
LRP	Living radical polymerizations
MADIX	Macromolecular design trough interchange of xanthates
MALDI	Matrix-assisted laser desorption/ionization time of flight
MMA	Methyl methacrylate
MS	Mass spectroscopy
NIPAM	N-Isopropylacrylamide
NMP	Nitroxide-mediated polymerization
PDI	Polydispersity index
PMMA	Poly(methyl methacrylate)
PRT	Primary radical termination
PS	Polystyrene
PVC	Poly(vinyl chloride)
RAFT	Reversible addition–fragmentation chain transfer polymerization
SET	Single electron transfer
TEMPO	2,2,6,6-Tetramethylpiperidinyloxy
THF	Tetrahydrofuran
TMI	1-(Isopropenylphenyl)-1,1-dimethylmethyl isocyanate
TPSE	1,1,2,2-Tetraphenyl-1,2-bis(trimethylsiloxy)ethane
VAc	Vinyl acetate
VDF	Vinylidene fluoride

1
Introduction

The definitions given for "telechelic oligomers" and "macromonomers" are not accurate and often lead to some confusion between these two terms in the literature. For instance, the term "macromonomer" is often replaced by "semitelechelic" [1]. Without prescribing any normalization, it is necessary to define well these two terms in order to correctly review the work done in both areas. To simplify, the definitions will be based on the functionality. Hence, a functionality of 2 on one chain end will be related to macromonomers. This comprises molecules bearing either a double bond or two polycondens-

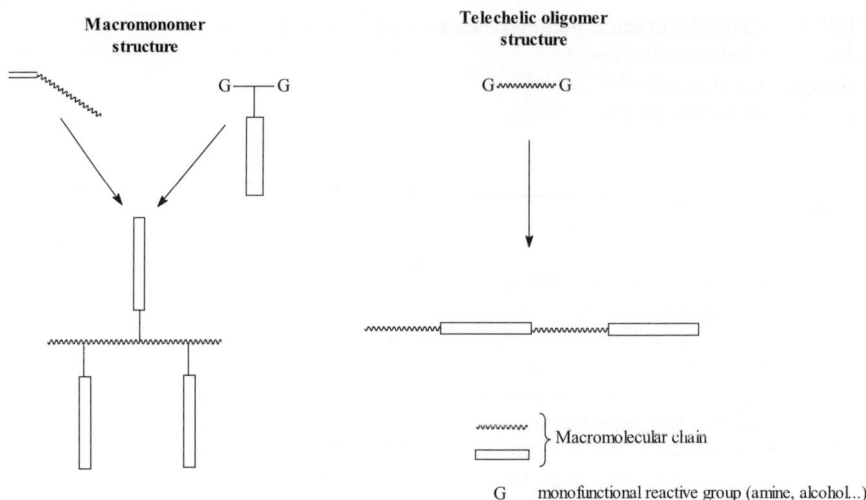

Scheme 1 The telechelic and macromonomer structures

able groups at the same chain end. The two polycondensable groups will be identical and named G, according to Scheme 1.

On the other hand, a functionality of 1 on each chain end will be related to telechelic compounds (Scheme 1). This includes diols, diamines, and diacides. Of course it also comprises diolefin compounds that usually lead to gels or networks. We can also note that when the G and G′ functional groups are different at each chain end, the appropriate term becomes heterotelechelic (Table 1). It is also necessary to specify the particular case of macromolecules bearing a well-identified G functionality at one chain end and a thermally reactivated group at the G′ chain end. These groups can be nitroxides, an iodine atom, xanthate, etc.and are commonly used in living radical polymerizations (LRP). These compounds may be classified as monofunctional oligomers (Table 1).

According to these definitions, macromonomers can be considered as precursors of graft copolymers, whereas telechelic oligomers will lead to multiblock copolymers. Among various methods for preparing telechelic oligomers and macromonomers, the radical polymerizations are certainly the most studied techniques. The success of radical polymerization may be due to the fact that no purification of the reactants is required and also the experimental conditions are generally not drastic. Furthermore, almost all the vinyl monomers can react through radical polymerization. In the 1990s, the syntheses of telechelic oligomers and of macromonomers were reviewed, for instance by Boutevin [2] and Rempp and Franta [3] and by Ito and Kawaguchi [4–6], respectively. These surveys deal with the use of conventional radical polymerizations, such as telomerization or dead-end polymerization (DEP), to achieve the telechelic or macromonomer structure.

Table 1 Schematic representations of each structural family

Structural family	Functionality by chain end
Telechelic oligomers	G〰〰〰〰G G functional group
Heterotelechelic oligomers	G〰〰〰〰G' G ≠ G; G and G' functional groups
Monofunctional oligomers	G〰〰〰〰G' G' = thermally reactivated group
Macromonomers with two polycondensable groups	G │ 〰〰〰〰 G G functional group
Macromonomers with a vinyl group	=〰〰〰

However, in 1994 LRP officially appeared that allowed for a major evolution of both the telechelic oligomers and the macromonomers. Indeed, LRP techniques provided new end groups such as nitroxides, dithioesters, xanthates, or even halogens, which could be easily modified.

The scope of this review is to consider all the radical techniques leading to a telechelic or a macromonomer structure. To this aim, it will be necessary to illustrate each method with some examples. We also emphasize that the number of studies in this area is numerous and some will be not considered. Indeed, this paper is not an index of publications on the synthesis of telechelic oligomers and macromonomers but more a comprehensive paper on how to achieve such structures by a radical process. Thus, this paper is divided in two distinct parts: synthesis of telechelic oligomers in the first part and synthesis of macromonomers in the second part. In the second part, we will also outline the new design of macromonomers.

2
Synthesis of Telechelic Oligomers by Radical Techniques

In the review paper of Boutevin [2], the different conventional radical polymerizations, such as telomerization or DEP, were mentioned to lead to a telechelic structure. Since 1990 these radical techniques have shown some progress in the synthesis of telechelic oligomers. We can remark that recent investigations into the understanding of the kinetics of the DEP of styrene are interesting. But the main breakthrough since 1990 concerning controlled radical polymerization (CRP) was realized by the occurrence of addition–fragmentation. For 10 years, addition–fragmentation transfer polymerization (AFT) has been used in lots of works in the area of telechelic oligomers.

Then, in 1994 LRP officially appeared and allowed for a major evolution of the telechelic oligomers. Indeed, LRP techniques were able to provide new end groups such as nitroxides, dithioesters, xanthates, or even halogens, which could be easily modified to lead to multiblock copolymers. The most efficient LRP methods are nitroxide-mediated polymerization (NMP) using alkoxyamines [7], atom transfer radical polymerization (ATRP) using alkyl halides [8], reversible addition–fragmentation transfer polymerization (RAFT) using thioesters [9], and iodine transfer polymerization (ITP) using alkyl iodides [10, 11].

For more than 10 years all these LRP techniques have provided numerous studies on the synthesis of telechelic oligomers. We show the possibilities offered by each LRP technique to synthesize telechelic oligomers.

2.1
Telechelic Oligomers Obtained by Telomerization

The telomerization process is based on the use of a transfer agent to control both the molecular weight and the chain-end functionality. The mechanism was developed in previous work [12, 13]. For a long time, telomerization remained almost the only technique allowing for the synthesis of telechelic oligomers.

Since the introduction of LRP, only a few studies have been realized on the synthesis of telechelic oligomers by means of telomerization reaction. It is, however, interesting to give a summary of the studies realized in this area since 1990 because very attractive chain-end functionalities can be easily obtained.

2.1.1
Radical Addition

Monodispersed telechelic oligomers, especially diols or diamines, are very interesting, because they can be used as antiwear and antitear additives for metallic surfaces, greasiness grade improvers for hydrocarbons, terpolymers for paints, and in the preparation of polyester and polyurethanes. The preparation of telechelic oligomers follows two synthetic strategies: the monoaddition of a monofunctional telogen onto a monofunctional taxogen or the addition of a monofunctional telogen onto a nonconjugated diene (Scheme 2). Some examples were given in the previous review. Here, two recent works are described.

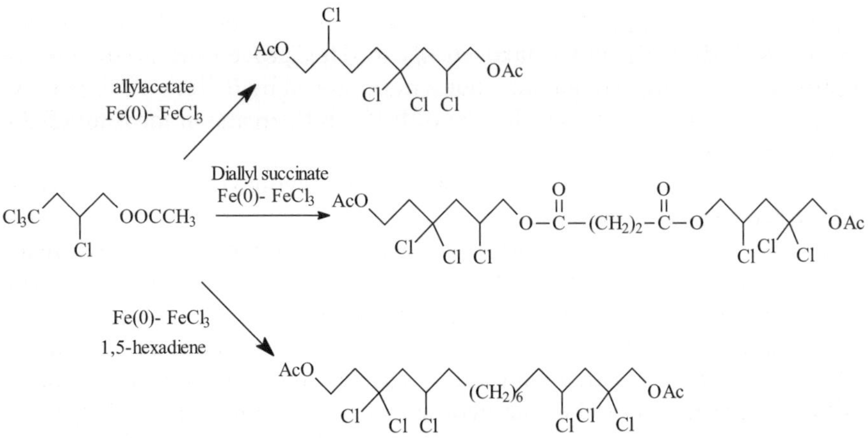

Telechelic Oligomers and Macromonomers by Radical Techniques 37

Scheme 2 Functionalization of 2,4,4,4-tetrachlorobutyl acetate

Table 2 Summary of telechelic oligomer obtained by radical addition (redox telomerization)

Taxogen (mmol/mmol TCEA)	Product	Fe(0)	Conv. (%)	Yield (%)
Allylacetate[a] (32/48)	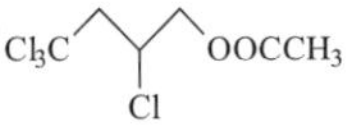	Turning or filing	100	93
Allylacetate (32/40)	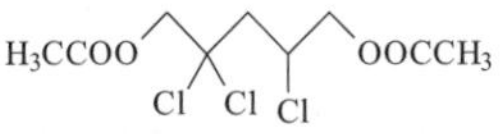	Turning	97	93
Allylacetate (32/40)		Filing	100	93
Ethyl ω-undecenoate (32/48)	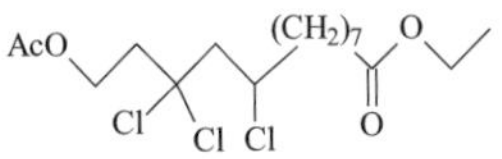	Filing or turning	92	93
Diallyl succinate (16/48)	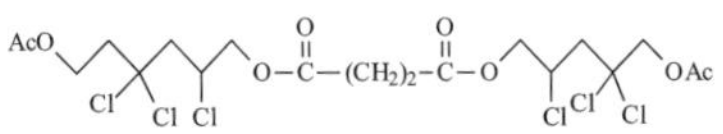	Filing	100	70
Metallylacetate (32/48)		Filing or turning	95	86
1,5-Hexadiene (72/24)	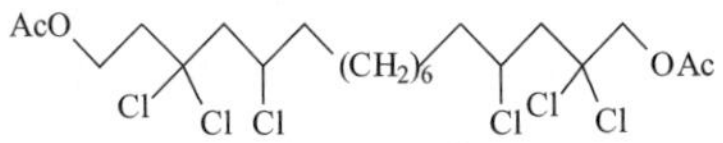	Filing or turning	–	89

TCEA 2,4,4,4-tetrachlorobutyl acetate
[a] Taxogen CCl_4

The first method is the use of a redox telomerization with different telogens, such as CCl$_4$ and a taxogen agent. This process was used for the synthesis of telechelic oligomers, but was improved by Bellesia et al. [14, 15]. These authors proposed the addition of Fe(0) in the reaction medium (FeCl$_3$ and benzoline) for promoting the Kharash addition of CCl$_4$ onto allyl acetate or methyl acetate (taxogen agent) in dimethylformamide (DMF) (with a 1 : 2 Fe(0) to FeCl$_3$ ratio) (Scheme 2). The combination of Fe(0) and FeCl$_3$ allows for the selectivity transformation of allyl acetate or methyl acetate at 80 °C to obtain trichloroethanol acetate (Table 2). A new reaction is performed in the same conditions with trichloroacetate as the telogen and allyl acetate as the taxogen at 100 °C. Thus, the reaction allows for the synthesis of telechelic oligomers with a good yield. The selective deprotection of acetate allows telechelic diols to be obtained.

The second method is a radical addition [16–18] of dithiols in the presence of *tert*-butyl peroxypivalate (Scheme 3) onto 1-undecenol at 75 °C for 5 h. This reaction leads to the synthesis of a diol in one step with a good yield (80%).

$$HS-(CH_2)_2-X-(CH_2)_2-SH \ + \ CH_2=CH-(CH_2)_9-OH$$

$$(H_3C)_3COOC(O)C(CH_3)_3 \ \Big| \ CH_3CN, \ 5 \ h, \ 75 \ °C$$

$$HS-(CH_2)_2-X-(CH_2)_2-S-(CH_2)_{11}-OH$$

Scheme 3 Monodisperse synthesis of a diol by radical addition performed at 75 °C with *tert*-butyl peroxypivalate

The presence of heteroatoms in the aliphatic chain (X is O or S) allows the melting point of the products to be reduced and also increases their solubility [19].

2.1.2
Nucleophilic Addition

Boyer et al.(C. Boyer, J.J. Robin, B. Boutevin, unpublished results) used the nucleophilic addition of thiolate onto the double bond of alkyl (meth)acrylate to obtain monodispersed telechelic oligomers. This method is based on the nucleophilic character of the thiolate ion in the presence of a monomer carrying two acrylate or methacrylate functions to obtain the corresponding telechelic oligomers. The nucleophilic addition of the thiolate ion onto the double bond is quantitative (Scheme 4).

The reaction is carried out in acetonitrile, in the presence of triethylamine in stoichiometric quantity with the thiol compound. The introduction of triethylamine allows the thiol–thiolate balance to be changed to give the

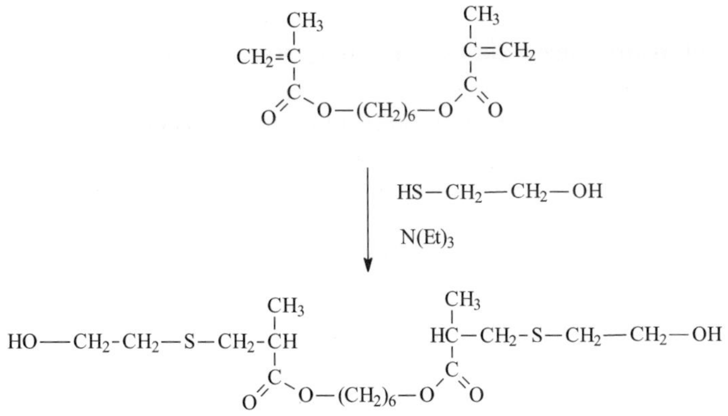

Scheme 4 Nucleophilic reaction for the addition of 2-mercaptoethanol onto 1,6-dimethacrylate hexane catalyzed by triethylamine ($N(Et)_3$)

thiolate ion, which can be added by Michael-type reaction onto the acrylic or methacrylic double bond. This reaction is quasi-instantaneous for acrylates, whereas the complete consumption of the methacrylate double bond requires 6-h reaction at 60 °C. In both cases, this reaction leads to monodispersed telechelic oligomers. The purification of hydroxy-telechelic oligomers only consists of evaporating the solvent.

Thus, it is possible to use this reaction to obtain new macromolecules (diols or diamines) with low glass-transition temperature (T_g) and no crystalline phase. Earlier works suggested the synthesis of hydroxy-telechelic oligomers by radical addition of thiol onto dienes. But, the products obtained showed poor solubility in organic solvents and a high melting point. Our process, however, improves the properties of diol compounds (better solubility and decrease of the melting point).

2.1.3
Chain-End Chemical Modification

The use of thiol compounds as transfer agents exclusively leads to monofunctional oligomers. To achieve bifunctionality requires a chain-end modification. Fock et al. [20–23] developed a new method based on the chemical chain-end modification of polymethacrylate telomers. These were previously obtained by telomerization reaction in the presence of mercaptoethanol or thioglycolic acid as transfer agents. The chemical modification can be summarized in two strategies:

1. Selective saponification leading to hydroxy-telechelic oligomethacrylates.
2. Selective transesterification of the chain-end ester groups leading to carboxy-telechelic oligomethacrylates.

2.1.3.1
Synthesis of Hydroxy-telechelic Oligomethacrylates

Firstly, the telomerization reaction of n-butyl methacrylate (Scheme 5, step 1) is realized in the presence of 2-mercaptoethanol [24–26]. The monofunctional oligomers obtained present a molecular weight of about 1250 with a functionality close to 1.

In a second step, the transesterification of the terminal ester group is possible by using 3-methylpentanediol. Owing to a steric effect, the reaction becomes selective, and it is possible to obtain a functionality very close to 2. The telechelic structure was evidenced by matrix-assisted laser desorption/ionization time-of-flight (MALDI-TOF) analysis. The telechelic oligomers were then reacted with 1,6-hexanediisocyanate to increase the molecular weight of the oligomers.

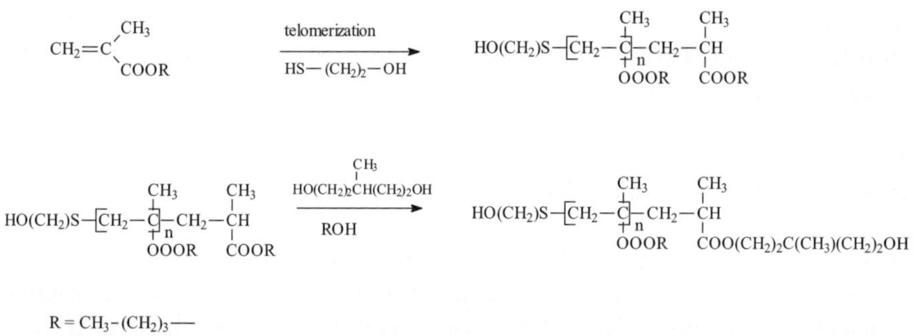

Scheme 5 Synthesis of hydroxy-telechelic oligomethacrylates by chemical modification (esterification)

2.1.3.2
Synthesis of Carboxy-telechelic Oligomethacrylates

The saponification of the terminal ester group allows the preparation of carboxy-telechelic oligomethacrylates [27]; hence, the terminal ester group is expected to react more than the ester homologues in the chain [24, 28]. Firstly, the telomerization of n-butyl methacrylate is realized in the presence of thioglycolic acid to get the acid group at the chain end. 4,4′-Azobis(4-cyanovaleric acid) (ACVA) was used as the initiator, to ensure a functionality of 1. The acid functionality was confirmed by MALDI-TOF analysis. Then the saponification of the terminal ester group was carried out. To avoid the saponification of the other ester groups, the the experimental conditions were changed (Table 3) and the acid functionality checked. It was shown that a complete saponification of the terminal ester group requires 4 equiv of KOH by ester function over 12 h in a dioxane/H_2O (2.5% v/v) solution.

Table 3 Variation of the functionality f for the resultant product after saponifying for different times in different solvents for n-butyl methacrylate oligomer ($M_n = 1400 \text{ g mol}^{-1}$). Saponification in the presence of 4 equiv of KOH

Solvents [2.5% (v/v)]	Time (h)					
	0	2.5	6	8	10	12
Tetrahydrofuran/H$_2$O	1	1.33	1.50	1.71	1.72	–
Butanone/H$_2$O	1	–	1.54	1.76	1.76	–
Dioxane/H$_2$O	1	1.39	–	–	1.89	1.95–2.05

The acid functionality was obtained in the range 1.95–2.05 by acido-basic titration.

The oligomers were analyzed by MALDI-TOF and different populations were characterized: telechelic oligomers were produced from direct initiation with ACVA and telechelic oligomers were produced from telomerization with thioglycolic acid.

2.2
Telechelic Oligomers Obtained by Dead-End Polymerization

DEP afforded the synthesis of telechelic oligomers [29]. With the telomerization process, DEP appeared to be the first free-radical polymerization leading to telechelic oligomers. Tobolsky [30] first stated the conditions of DEP, i.e., the half life of the growing species has to be equivalent to that of the initiator. These specific conditions result in an unusual high rate of termination, which allows for the synthesis of oligomers. The telechelic structure will be obtained by combining a termination mode exclusively by recombination with the use of a difunctional initiator.

Boutevin [2], in his review on "Telechelic oligomers by radical reactions" developed the categories of functional initiators, i.e., diazoic compounds, hydrogen peroxide, and oxygenated substances. He examined the different reactivities and combinations of such initiators with monomers in order to synthesize telechelic oligomers. Boutevin [2] also summarized the monomers able to totally recombine or to avoid termination by disproportionation. He showed a quantitative amount of recombination only for styrene, acrylates, dienes, and acrylonitrile [31–33].

We examine the recent developments made in DEP conditions with monomers such as styrene, acrylates, fluoro-type monomers, and N-isopropylacrylamide, with the aim of synthesizing telechelic oligomers.

2.2.1
Styrene

2.2.1.1
Synthesis of Telechelic Oligostyrene

Styrene is probably the most used monomer in DEP conditions [34]. Recently, some authors developed the synthesis of carboxy-telechelic polystyrene (PS) through DEP [35–37] by using ACVA. In a recent publication [38], we focused on developing the synthesis of carboxy-telechelic PS by improving the experimental conditions of DEP (Scheme 6).

By varying the experimental conditions, mainly the initial initiator concentration, we ended up with oligomers exhibiting molecular weights in the range 1500–25 000. The purification of such oligomers was investigated in detail. By-products produced from recombination of ACVA radicals [39] were eliminated by water extraction. The bifunctionality (f_{COOH}) of oligomers was proved, as shown in Table 4.

Finally, the bifunctionality was investigated through a MALDI-TOF analysis that perfectly characterized a series of four isotopic peaks. The major one corresponds to the expected telechelic structure. Two other series were shown

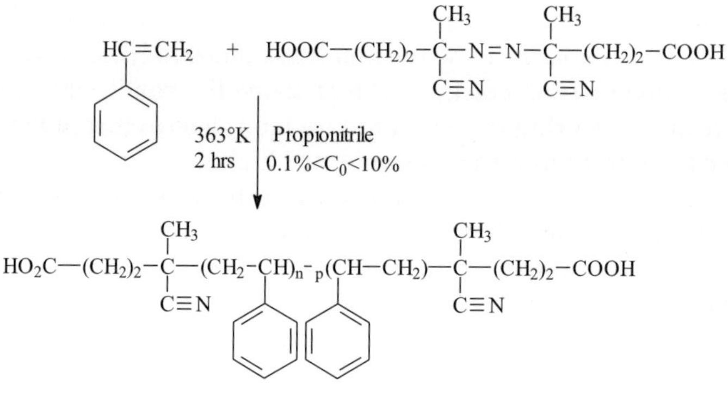

Scheme 6 Synthesis of carboxy-telechelic polystyrene by dead-end polymerization (*DEP*)

Table 4 Acid functionality f_{COOH} of oligostyrene obtained by dead-end polymerization (*DEP*) with 4,4′-azobis(4-cyanovaleric acid) (*ACVA*)

	$C_0 = [I_2]_0/[M]_0$ (%)	^{1}H NMR	Conductimetric titration
f_{COOH}	10	1.96	1.85
	5	1.90	1.90

to be cationized acid compounds. The last peak, however, shows a monofunctional oligomer with a double bond at the chain end. These last oligomers might result in a disproportionation as a termination mode. The authors raised the fact that if disproportionation occurred, the saturated homologue to HOOC–PS$^=$ should be visible in MALDI-TOF analysis. In the expected region of the saturated monofunctional oligomer, no peak was visible. Actually, HOOC–PS$^=$ was produced by occurrence of fragmentation during ionization [40].

2.2.1.2
Kinetics Approach

The synthesis of carboxy-telechelic PS by DEP is now well established. But the major breakthrough in this synthesis certainly concerns the kinetics approach and the prediction of the cumulative degree of polymerization.

The kinetics aspect [41] of the reaction shown in Scheme 6 was studied and an entire mechanism was proposed, including the different kinetics constants. This new mechanism shows the occurrence of a "new" termination reaction by recombination, namely, primary radical termination (PRT), characterized by a kinetics constant k_{PRT}. PRT consists of a reaction between a primary radical (directly produced by the initiator) with a growing radical. The "conventional" bimolecular recombination was also represented by its kinetics constant k_{tc}.

We used and developed some kinetics relationships in order to evaluate such recombination kinetics constants (Table 5). The different models presented in Table 5 directly give the ratio $k_{PRT}/k_i k_p k_{PRT}$, using k_i of α,α'-azobis(isobutyronitrile) (AIBN) as an approximation.

Table 5 shows homogeneous values whatever the chosen model equation. More importantly, it also shows a very high value for the constant of PRT. We can appreciate values of k_{PRT} around 10^{10} mol^{-1}s^{-1}l. This reaction is at least 100 times faster than that of bimolecular termination.

Table 5 Recombination termination kinetics constants for the DEP of styrene with ACVA

Model equations	$k_{prt}/k_i k_p$ (mol s l^{-1})	$k_{prt} 10^{10}$ (mol^{-1}s^{-1}l)	$k_{tc} 10^7$ (mol s l^{-1})
Bamford et al. [349]	5464	1	7.2
Olaj [350]	5300	1	7.3
Deb and Meyerhoff [351], Mahabadi and Meyerhoff [352]	6800	1.4	–
Ito [353]	3179	0.63	9

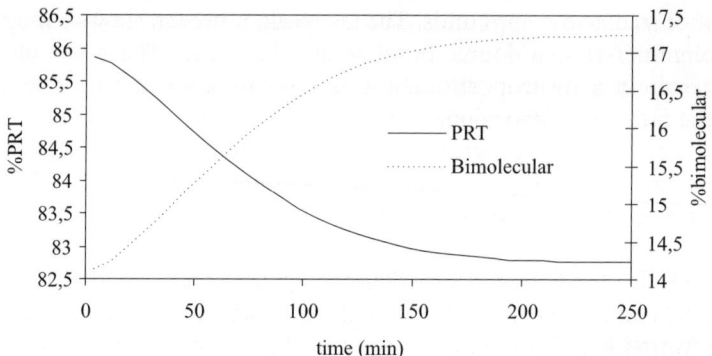

Fig. 1 Predictions of the percentage of primary radical termination (*PRT*) and bimolecular termination for the dead-end polymerization of styrene with [ACVA]$_0$/[Sty]$_0$ = 0.1. *ACVA* 4,4′-azobis(4-cyanovaleric acid), *Sty* styrene

The next step consisted in showing the influence of PRT compared with that of conventional bimolecular termination in conditions of DEP, i.e., with a high initial initiator concentration. This was achieved by simulation software (Fig. 1). We proved that PRT is the preponderant recombination reaction, allowing for the synthesis of low molecular weight telechelic PS.

A theoretical approach to the cumulative degree of polymerization was finally undertaken in the conditions of DEP. However, no kinetics model [12] considers the PRT, which takes place preponderantly in the case of DEP. We [42] then developed a new kinetics relationship connecting the degree of polymerization with the kinetics constants of both PRT and bimolecular termination (Eq. 1):

$$\left(\overline{DP_n}\right)^2_{\text{inst}} = (1+a)^2 \times \frac{k_p^2}{k_{tc}} \times \frac{1}{2fk_d[I_2]} \times \left[[M]^2 - \left(\frac{k_{\text{PRT}}}{k_i k_p}\right)^2 \times \frac{Rp^2}{[M]^2} \right]. \quad (1)$$

The experimental calculation of both $k_{\text{PRT}}/k_i k_p$ and k_p^2/k_{tc} allows for the determination of theoretical $\left(\overline{DP_n}\right)_{\text{inst}}$, leading to $\left(\overline{DP_n}\right)_{\text{cum}}$ by iteration. We showed that our model can be extended to a wide range of $\left(\overline{DP_n}\right)_{\text{cum}}$ from 10 to 150.

2.2.2
Acrylates

Acrylates are of particular interest for the synthesis of telechelic oligomers through the technique of DEP. Actually, acrylates are well known to give only recombination as a termination reaction. Also, for the aim of obtaining block copolymers by polycondensation, telechelic oligoacrylate may represent a new soft segment [43].

In this section we will present the recent developments concerning first acrylates and second fluoroacrylates.

2.2.2.1
Synthesis of Telechelic Oligoacrylate

Banthia et al. [44] first performed the polymerization of ethyl hexyl acrylate in DEP conditions (Table 6) but without using any solvent. The calculated carboxy functionalities were around 2 and the molecular weights were about 10^4. The number-average molar masses (M_n) were unexpectedly high for DEP conditions and were probably due to the bulk conditions.

We recently performed [45] this reaction using a solvent, i.e., propionitrile. We observed a lowering of the molecular weight but functionalities dropped to almost one acid group per chain. This result was correlated to a high amount of transfer reaction of the growing radical mainly to the solvent. Hence, changing from propionitrile to methyl-2-propanol was responsible for an increase of the functionality (about 1.7) for M_n of about 5000 g mol^{-1}. In order to lower the reactivity of the poly(alkyl acrylate) growing radical, we performed radical polymerization of acrylate at low temperature using a redox system (dimethylaniline-catalyzed benzyl peroxide). At $-20\,°C$ the conversion profile for both the initiator (benzyl peroxide) and the acrylate looked like a conventional conversion profile of DEP, i.e., the initiator being consumed faster than the monomer. However, the molecular weights obtained were still quite high (about 10^4).

This low-temperature initiation probably opens the way to the synthesis of new telechelic acrylates by choosing the right peroxide/redox system.

Table 6 Synthesis of carboxy-telechelic ethyl hexyl acrylate in DEP conditions with ACVA

$C_0 = [I_2]_0/[M]_0$ (%)	T (°C)	M_n (g mol^{-1})	f_{COOH}
6.5	110	10 200	2.07
6.5	100	11 500	2.01
6.5	90	12 300	2.02
6.5	80	15 000	2.04
8.2	100	10 200	1.99
9.8	100	10 100	2.01
13.1	100	9500	2.03

2.2.2.2
Synthesis of Telechelic Oligofluoroacrylate

Attempts were made at synthesizing telechelic oligomers of fluoroacrylate monomers. These polymers have great potential in coatings or for optical materials [46]. Like acrylates, fluoroacrylates can be good candidates for DEP as they exhibit only termination by recombination. Radical polymerizations have been performed in DEP conditions for 2,2,2-trifluoroethyl α-fluoroacrylate (FATRIFE) initiated by *tert*-butylcyclohexyl peroxydicarbonate at 75 °C [47]. However, despite a high initial initiator concentration, high molecular weights for poly(FATRIFE) were obtained (Table 7).

Table 7 Evolution of number-average degree of polymerization (DP_n) with initial initiator concentration for the radical polymerization of 2,2,2-trifluoroethyl α-fluoroacrylate with *tert*-butylcyclohexyl peroxydicarbonate at 75 °C in acetonitrile

$[M]_0$ (mol l^{-1})	$[I_2]_0$ (mol l^{-1})	Time (min)	α_M [a] (%)	(DP_n) [b]
0.584	0.057	30	1	41
0.587	0.026	60	1	48
0.582	0.006	60	1	130

[a] Monomer conversion [b] Determined by ^{1}H NMR

The development of kinetics relationships (see previously for styrene) allowed us to demonstrate that the PRT, responsible for low M_n, was not favored at all when α-fluoroacrylate was used in DEP. The $k_{prt}/k_i k_p$ ratio was calculated to be about 20, whereas for styrene the ratio was calculated to be about 6×10^3. We showed that the resulting amount of PRT was about 40% for FATRIFE and about 85% (Fig. 1) for styrene for the same C_0.

The synthesis of telechelic α-fluoroacrylate with low molecular weight has not been achieved so far by using the technique of DEP.

2.2.3
Fluoro-type Monomers

Recent investigations were made into the synthesis of telechelic oligomers by the technique of DEP for monomers such as vinylidene fluoride (VDF) or hexafluoropropene. Saint-Loup et al. [48, 49] described the efficiency of these monomers in conventional radical polymerization through DEP conditions. First the photopolymerization of VDF was investigated with hydrogen peroxide, leading to original hydroxycarboxy telechelic poly(vinylidene fluoride) with M_n ranging from 400 to 4000 g mol^{-1}.

The synthesis of carboxyl end groups was explained by the fact that OH radicals may add onto the CF_2, leading to intermediate $FOCCH_2$. In the presence of water, this entity will become $HOOCCH_2$ that will initiate a new chain and end up in a carboxyl end group.

Following the same idea, dead-end copolymerization of VDF with hexafluoropropene was realized with hydrogen peroxide [48]. Once again, oligomers with M_n ranging from 700 to $3500\,\text{g mol}^{-1}$ were obtained with satisfactory conversion rates. More interestingly, the authors assessed a carboxyl functionality of about 1.85, indicating that addition of OH radical onto CF_2 or CF becomes the major pathway. The authors also revealed the presence of an inner unsaturated $CH=CF$ bond, induced by HF elimination according to the experimental conditions. Finally, the authors performed a reduction of carboxylic groups onto hydroxyl groups in the presence of $LiAlH_4$ (Scheme 7). These new telechelic oligomers can be of great interest for further polycondensation reactions.

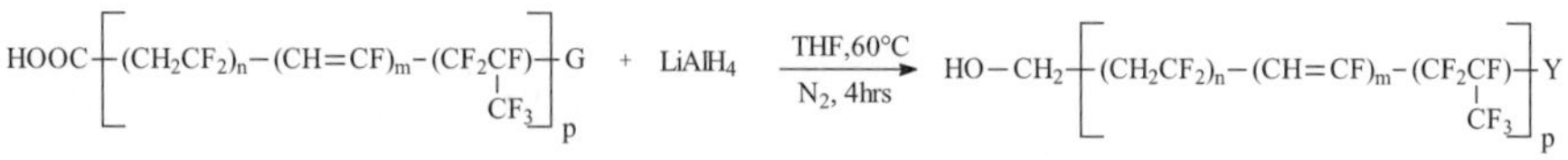

$$HOOC-[(CH_2CF_2)_n-(CH=CF)_m-(CF_2CF)\!\!-\!\!\underset{CF_3}{|}]_p-G \;+\; LiAlH_4 \xrightarrow[N_2,\,4hrs]{THF,\,60°C} HO-CH_2-[(CH_2CF_2)_n-(CH=CF)_m-(CF_2CF)\!\!-\!\!\underset{CF_3}{|}]_p-Y$$

G represents a hydroxyl, hydroxymethyl or carboxylic function
Y represents a hydroxyl or hydroxymethyl function

Scheme 7 Fluoro-telechelic macrodiols induced by DEP followed by reduction

2.2.4
Other Monomers

Different monomers, such as N-isopropylacrylamide (NIPAM), were tested in DEP conditions. Smithenry et al. [50] developed the synthesis of carboxy-telechelic poly(NIPAM) using the concept of DEP (Scheme 7). The values of M_n were estimated to be in the range 5×10^3–$32\times10^3\,\text{g mol}^{-1}$, depending on the concentration of the initiator. A kinetics model was also proposed and good agreements between predicted and experimental molecular weights were shown. It is interesting to note that aggregates of poly(NIPAM) were evidenced by light scattering. The authors showed that aggregation became irreversible even with addition of salt.

2.3
Telechelic Oligomers Obtained by Addition–Fragmentation

2.3.1
Use of Chain Transfer Agents in Addition–Fragmentation

Among several radical techniques, the free-radical addition-fragmentation chain transfer reaction appears to be an unrivaled method for the synthesis of molar-mass-controlled heterotelechelic polymers from a wide range of vinylic-type monomers. The addition–fragmentation processes were first studied in radical organic chemistry, and occur whenever a growing macroradical reacts with a reagent bearing both an activated double bond and a weak linkage located somewhere else on the molecule. Such a process was also identified as an effective means for controlling the molar mass of vinyl polymers, avoiding the use of conventional chain transfer agents (CTAs) based on thio derivatives.

The mechanism of the addition–fragmentation process is rather complex (Scheme 8) [51–53] as it involves different steps: addition of the macroradical onto the CTA followed by a subsequent β-fragmentation, but also an intramolecular substitution on a peroxydic bond may occur, depending on the CTA structure. The overall mechanism of addition–fragmentation is often more complicated than shown in Scheme 8 [54] and was studied especially in terms of driving forces in free-radical addition–fragmentation [55]. Colombani and Chaumont [56] presented a general review in which they mainly focused on recent developments in the large area of addition–fragmentation. More recently, CTAs were even involved in emulsion polymerization [57].

This process is schematically identical to a classic atom transfer reaction because of the termination of the polymer chain and the reinitiation of a new one. Thus, as a first approach, it is possible to apply the classical Mayo equa-

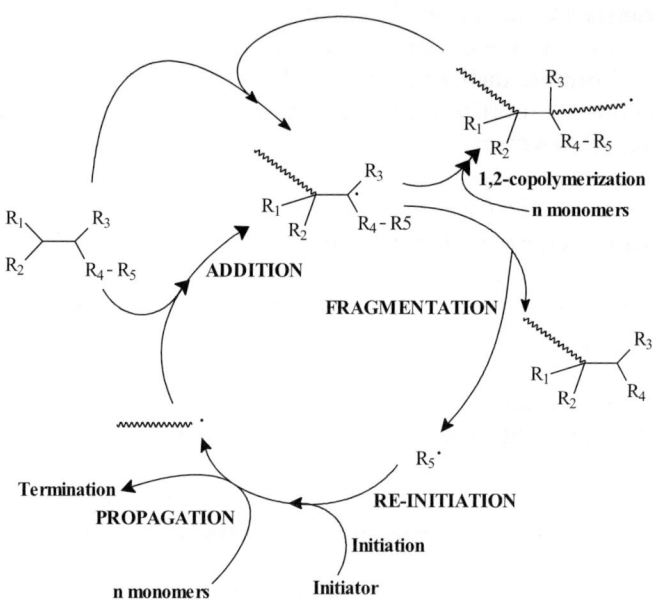

Scheme 8 Free-radical addition–fragmentation processes

tion to calculate the chain transfer constant ($C_{tr,CTA} = k_{tr}/k_p$) (Eq. 2):

$$\frac{1}{DP_n} = \frac{1}{DP_{n,0}} + C_{tr,CTA}\frac{[CTA]}{[M]} \,. \tag{2}$$

The CTAs which follow the addition–fragmentation [58, 59] mechanism are of particular interest in organic and polymer chemistry. Recently, many studies have shown that allyl, acrylyl, and allenyl transfers to alkyl halides represent powerful synthetic tools to prepare sophisticated molecules.

In this review we will focus on how the addition–fragmentation sequence can provide controlled functionality at the end of the polymer chains. An attractive feature of this technique is also the concomitant incorporation of a terminal functional group following fragmentation, the functional group being vinylic (allyl, acrylyl, allenyl, etc.), ketonic, carboxylic, amino, halogeno, epoxidic, etc., depending on the system. Nowadays, most studies are conducted to design new CTAs in order to control the functionality of both chain ends of the telechelic polymer.

Two distinct sites of the CTAs are involved in the addition step and the fragmentation step [54]. Thus, it is theoretically possible to design each site separately, in order to (1) control the reactivity of the CTA, i.e., the chain transfer constant value, which is mainly influenced by the nature of the addition site, and (2) control the nature of the α-functional group, which is mainly influenced by the evolution of the fragmentation site. For the design of this latter site, the reactions involved in the fragmentation process generally deal with two classic reactions studied in organic chemistry: the β-scission reaction and the intramolecular homolytic substitution called SHi.

The preparation of such CTAs involves a nonnegligible part of organic chemistry. The functional end groups of the oligomers obtained may be further modified into others by classic reactions to extend the potentially available end fragments.

The general form [56] for CTAs involved in addition–fragmentation is CX = C(Y) – W – G (Scheme 9). However, CTAs can potentially be separated in three distinct types, A, B, and C, as shown in Scheme 9, leading to three types of α,ω-difunctional oligomers.

Colombani and Chaumont [54] comprehensively summarized all the studies concerning the synthesis of α,ω-difunctional oligomers through addition–fragmentation by using such transfer agents. In this chapter we only present some examples of studies using A, B, or C types of CTAs. Table 8 shows some CTAs involved in addition–fragmentation, leading to the expected telechelic structure.

Few monomers have been studied in addition–fragmentation polymerization. Mainly styrene, acrylate, and methacrylates have been used so far in addition–fragmentation to obtain telechelic oligomers. As an example, styrene and methyl methacrylate (MMA) [60, 61] were polymerized through an addition–fragmentation process, using allylic sulfides as CTAs (entries 12

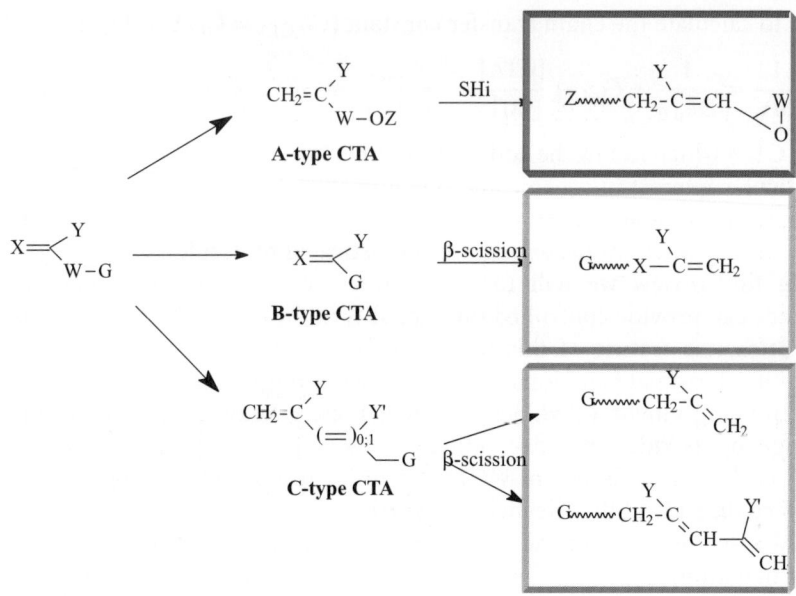

Scheme 9 Synthesis of α,ω-difunctional oligomers through addition-fragmentation processes. *CTA* chain transfer agent

and 14 in Table 9). Meijs et al. [60] used a combination of G and Y functions for the C-type CTA (Scheme 9) in order to obtain telechelic oligomers potentially available for further polycondensation reactions. Polymerizations using such CTAs are sumarized in Table 9. These experiments carried out at 60 °C showed a very low conversion whatever the chosen CTA. The authors also obtained chain transfer constants in the range 0.3–1.9 depending of the CTA type towards monomers. More important is the result of the functionality obtained by ^{1}H NMR. Indeed, Meijs et al. proved that dihydroxy PS and poly(methyl methacrylate) (PMMA) can be obtained by addition–fragmentation with OH functionality close to 2. Phthalamido and COOH groups can be obtained as chain ends of PS and PMMA with very high efficiency (Table 9).

Another interesting survey is the use of A-type CTA allylic peroxides [62] to incorporate terminal functional groups. This type of CTA involves an intramolecular substitution, called SHi, which leads to an epoxidic end group on the polymer chain. The group attached onto the peroxide of the CTA, called Z in Scheme 9, gives the other functionality. The first CTAs synthesized were ethyl-2-[1-(1-n-butoxyethylperoxy)ethyl]propenoate [63], ethyl-2-[1-((2-tetrahydrofuranyl)peroxy)ethyl]propenoate (ETPEP) [64], and methyl 2-*tert*-butylperoxymethylpropenoate. The synthesis of these CTAs is rather complex and leads to a Z end group not so interesting for further polycondensation reactions. For instance, the synthesis of ETPEP requires a two-step

Table 8 A, B, and C types of chain transfer agents (CTAs) involved in addition–fragmentation leading to a telechelic structure

CTAs	Entry	X	Y	W	Z	G	Y'	Refs.
A-type	1	–	CO_2Et	$CHCH_3$	OH	–	–	[354]
	2	–	CO_2Et	CH_2	OtBu	–	–	[355]
	3	–	CO_2Et	$CHCH_3$	$OSiMe_3$	–	–	[51]
	4	–	Ph	$CHOCH_3$	$OCMe_2Ph$	–	–	[356]
	5	–	CH_3	C=O	$OCMe_2Ph$	–	–	[356]
	6	–	CO_2Et	$CHCH_3$	⟨tetrahydrofuran⟩	–	–	[64]
B-type	7	CH_2	Ph	–	–	OCH_2Ph	–	[61]
	8	CH_2	CN	–	–	OCH_2Ph	–	[298]
	9	S	Ph	–	–	CH_2Ph	–	[357]
	10	CH_2	OMe	–	–	Me	–	[358]
	11	CH_2	OCH_2-p-HOPh	–	–	OCH_2-p-HOPh	–	[358]
C-type	12	–	CO_2Et	–	–	StBu	–	[60, 198]
	13	–	CO_2Me	–	–	Br	–	[298]
	14	–	CO_2H	–	–	SCH_2CO_2H	–	[60]
	15	–	CN	–	–	StBu	–	[61]
	16	–	H	–	–	Br	H	[359]
	17	–	H	–	–	C(SMe)CN	H	[198]
	18	–	CH_3	–	–	StBu	CO_2Me	[359]
	19	–	CO_2Ph	–	–	$C(Me)_2Ph$	–	[360]

Table 9 Number-average molar mass, conversion, chain transfer constant, and functionality (f) for polymerizations carried out in the presence of allylic sulfide CTAs

CTAs	Monomer	Conversion (%)	M_n (g mol^{-1})	C_{tr}	f	Type of functionality
CH$_2$=C(CH$_2$-S-CH$_2$CH$_2$OH)(COOH)	Sty	1.5	6100	1.81	1.2	OH/COOH
	MMA	2.8	24 000	0.27	–	OH/COOH
CH$_2$=C(CH$_2$-S-CH$_2$CH$_2$OH)(COOCH$_2$CH$_2$OH)	Sty	2	5300	0.77	2.1	OH
	MMA	–	3350	0.4	1.4	CH$_2$OC(O)
	BA	10	5800	1.88	0.9	CH$_2$OC(O)
CH$_2$=C(CH$_2$-S-CH$_2$CH$_2$OH)(COOCH$_2$CH$_2$-Phtalamide)	Sty	–	6900	1.87	1.0	OH
	MMA	–	5900	0.72	1.2	Phthalamido

Sty styrene, *MMA* methyl methacrylate, *BA* butyl acrylate

reaction [65]: photooxygenation of ethyl tiglate is realized and the product obtained then reacts in a second step with 2,3-dihydrofuran in the presence of *p*-toluenesulfonic acid to give ETPEP with 60% yield. ETPEP was shown to regulate the molecular weight for monomers such as styrene, MMA, and *n*-butyl acrylate (*n*-BA). The bifunctionality was proved for the oligomers obtained to be formate and glycidic ester end groups, which are not easily polycondensable functions.

Using the same idea, Colombani et al. [51] synthesized peroxysilane CTAs. These CTAs showed good activity towards styrene, MMA and vinyl acetate (VAc). The resulting polymers carried a silyloxyl fragment at one end and a glycidic ester group at the other end.

2.3.2
Catalytic Chain Transfer

Undoubtedly the addition–fragmentation process is the nonliving radical polymerization that opens the route to new telechelic oligomers with good control of the molecular weight and good respect of the bifunctionality. However, accessing new CTAs is certainly the main limit of this technique. We showed that the synthesis of such CTAs is quite often very complex and involves many reaction steps. The investigation of catalytic chain transfer (CCT) for accessing new CTAs may allow further developments of addition–fragmentation.

The CCT technique is based upon the fact that certain Co(II) complexes such as cobaltoximes catalyze the chain transfer to monomer reaction. The mechanism is believed to consist of two consecutive steps [66] (Scheme 10). First, a growing polymeric radical R_n undergoes a hydrogen transfer reaction with the Co(II) LCo complex to form a polymer (or an oligomer) with a terminal double bond $P_n^=$ and the corresponding Co(III) hydride LCoH. Then, Co(III) hydride LCoH reacts with a monomer to produce both Co(II) and a monomeric radical. The mechanism of CCT is perfectly described by Gridnev et al. [67–69], who also proved that propagation in the presence of a cobalt catalyst occurs by a free-radical mechanism and not by a coordination mechanism. Barner-Kowollik et al. [70] also supported the mechanism of CCT by the use of MALDI-TOF analyses.

For instance, Haddleton et al. [71, 72] developed this technique to obtain telechelic PMMA. Unlike Hutson et al. [73], Haddelton et al. performed the CCT onto hydroxyethyl methacrylate (HEMA) and benzyl methacrylate (BMA). The HEMA dimer macromonomer and the BMA dimer macromonomer were first synthesized by CCT and purified. The HEMA dimer was then engaged in addition–fragmentation of MMA [71]. The authors first observed a lowering of the C_{tr} compared with the corresponding trimer of MMA. However, high conversions were obtained, up to 80% for M_n in the range 10^4–7×10^4 g mol^{-1}. The MALDI-TOF analysis proved the expected hydroxy-telechelic structure obtained through β-scission of the HEMA dimer macromonomer. Finally, they performed the addition–fragmentation of MMA in the presence of BMA dimer macromonomer [72]. Once again, conversions reached about 80% for M_n of about 2×10^4 g mol^{-1} and polydispersity index (PDI) around 2.5. The authors then performed hydrogenation of the synthesized oligomers to get carboxy-telechelic PMMA. The telechelic structure, confirmed by MALDI-TOF analysis, is as follows:

$$\text{Me-}\underset{\underset{HO}{\overset{C}{\diagdown}}\overset{}{\diagup}O}{\overset{\overset{Me}{|}}{C}}-(CH_2-\underset{\underset{CO_2Me}{|}}{\overset{\overset{Me}{|}}{C}})_x-CH_2-\underset{\underset{HO}{\diagup}}{\overset{\overset{CH_3}{|}}{CH}}\overset{}{\diagdown}C=O$$

2.4
Telechelic Oligomers Obtained by Other Conventional Radical Polymerizations

2.4.1
Use of Initer/Iniferter Systems

In conventional radical polymerization (CRP), the use of well-designed initiators gives various polymers or oligomers with controlled end groups [74, 75]. The concept of initer is given for compounds able to both initiate and terminate a polymerization [76]. Many workers [77–82] have been interested in synthesizing such compounds because they offer an easy way to get telechelic oligomers. The most promising initer compounds used in CRP to get telechelic structures are the 1,2-disubstituted tetraphenylethanes. The general structure is given in Scheme 10, which shows that 1,2-disubstituted tetraphenylethane also serves as a C–C-type thermal initer leading to a telechelic polymer, X being the potentially condensable group.

In Table 10 we have gathered different 1,2-disubstituted tetraphenylethanes reported in the literature to get telechelic polymers. We can remark that few studies were undertaken in the area of telechelic polymers; hence, despite a one-step reaction to get a telechelic structure, the main interest attributed to initer systems concerns the ability to restart a block copolymerization. The number of publications concerning the synthesis of diblock copolymers may prove this assumption. Under certain polymerization conditions, the chain ends, comprising the last monomer unit and the primary radical formed from the intiator, may split up into new radicals able to reinitiate further polymerization of a second monomer, leading to block copolymers. This is certainly the reason why 1,2-disubstituted tetraphenylethane does not present such interesting condensable functions (X in Scheme 10) for polycondensation reactions (Table 10).

The use of 1,2-disubstituted tetraphenylethanes is, however, of great interest because it allows for the synthesis of telechelic oligomers in a one-step reaction for monomers such as MMA which give a high amount of dispropor-

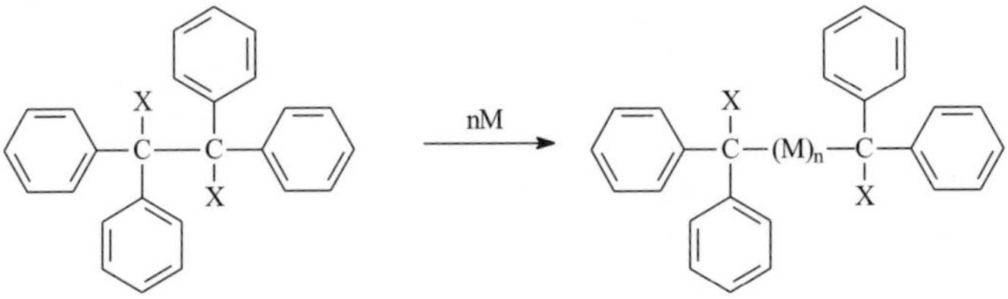

Scheme 10 General structure of 1,2-disubstituted tetraphenylethane

Table 10 Some 1,2-disubstituted tetraphenylethanes used to form telechelic oligomers

X group (Scheme 1)	Monomer(s)	M_n (g mol^{-1})	Refs.
CN	n-BMA/tert-BMA	5000/6000	[82]
	MMA	2500	[361, 362]
C_2H_5	MMA	–	[363, 364]
$O-C_6H_5$	MMA	–	[365–368]
$OSi(Me)_3$	MMA	13 000	[83]
$OSi(Me)_2C_2H_4CF_3$	2,2,2-trifluoroethyl methacrylate	18 000	[369]

BMA benzyl methacrylate

tionation in conventional radical polymerization. For instance, Roussel and Boutevin [83] used 1,1,2,2-tetraphenyl-1,2-bis(trimethylsiloxy)ethane (TPSE) in the polymerization of MMA (Scheme 11) at 80 °C. Scheme 11 shows that the diphenylmethyl radicals generated by TPSE are found to reversibly combine with the growing radicals, leading to telechelic PMMA. Scheme 11 also shows that a telechelic structure is obtained without the use of an additional "conventional" initiator, such as diazoic or peroxy compounds.

The authors characterized the expected structure by means of ^{1}H NMR. By performing kinetics analysis of the MMA polymerization, the authors observed an inhibition period of about 1 h, corresponding to the lifetime of the monoadduct formed.

Trimethylsilyl-terminated PMMA was well characterized and proved the potentiality of such a method. However, it could be interesting to chemically modify the end group of PMMA, aiming at further polycondensation reactions.

Focusing on telechelic polymers, the concept of "iniferter" is probably more interesting. Like initer, iniferter compounds will be able to initiate and terminate the polymerization. They also function as a CTA [84]. To get telechelic polymers by radical polymerization, it is necessary to use compounds with high transfer constants along with the radical initiator. Cho and Kim [85] suggested such a system, based on the use of two compounds bear-

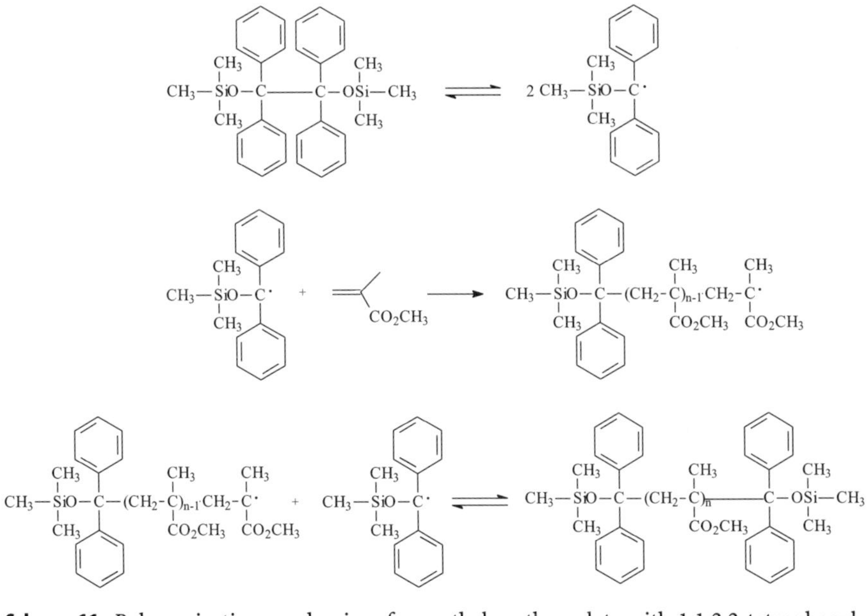

Scheme 11 Polymerization mechanism for methyl methacrylate with 1,1,2,2-tetraphenyl-1,2-bis(trimethylsiloxy)ethane

ing the same functional group. The first compound will be an initiator and the second one will act as a transfer agent. Cho and Kim [85] performed the radical polymerization of vinyl monomers in the presence of both 4,4'-azobis(cyanopropanol) and allyl alcohol (Scheme 12).

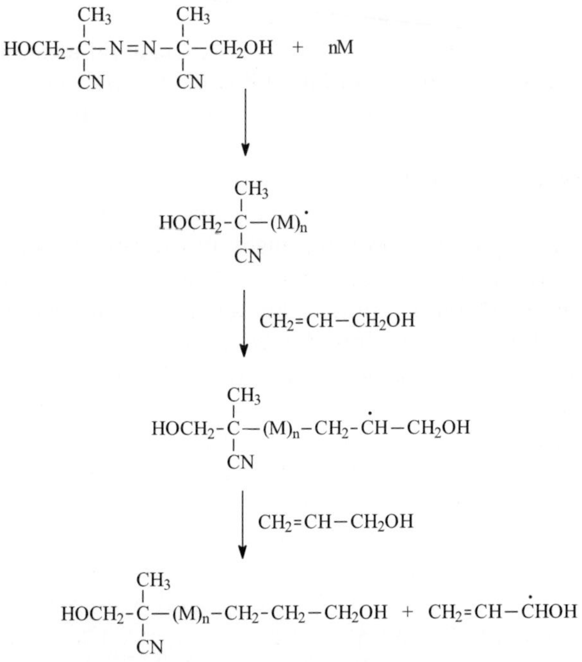

$$HOCH_2-\underset{\underset{CN}{|}}{\overset{\overset{CH_3}{|}}{C}}-N=N-\underset{\underset{CN}{|}}{\overset{\overset{CH_3}{|}}{C}}-CH_2OH \;+\; nM$$

$$\downarrow$$

$$HOCH_2-\underset{\underset{CN}{|}}{\overset{\overset{CH_3}{|}}{C}}-(M)_n^{\bullet}$$

$$\downarrow CH_2=CH-CH_2OH$$

$$HOCH_2-\underset{\underset{CN}{|}}{\overset{\overset{CH_3}{|}}{C}}-(M)_n-CH_2-\overset{\bullet}{C}H-CH_2OH$$

$$\downarrow CH_2=CH-CH_2OH$$

$$HOCH_2-\underset{\underset{CN}{|}}{\overset{\overset{CH_3}{|}}{C}}-(M)_n-CH_2-CH_2-CH_2OH \;+\; CH_2=CH-\overset{\bullet}{C}HOH$$

Scheme 12 Synthesis of hydroxy-telechelic polymers by using an iniferter system

Table 11 Radical polymerizations of vinyl monomers in the presence of 4,4'-azobis(cyanopropanol) and allyl alcohol (AA)

Monomers	[AA] (%)	Conversion (%)	M_n (PDI)	f_{OH}/chain[a]
Styrene	0.95	83	4300 (1.7)	2.01
	1.9	68	3600 (1.8)	2.03
Vinyl acetate	0.23	82	2400 (1.9)	2.21
	0.45	53	1900 (2.1)	2.28
MMA	0.77	84	5700 (1.9)	1.95
	1.54	74	4900 (2.0)	2.03
n-BA	0.48	91	5400 (1.9)	2.15
	0.96	70	4200 (1.9)	2.19

[a] Calculated from gel permeation chromatography (polystyrene standards)

Allyl alcohol, acting as a transfer agent, allows the terminal hydroxyl function to be obtained. The chain transfer constant of allyl alcohol was calculated to be about 2×10^{-2} towards poly(styryl radical). The authors used different monomers (Table 11) and always got functionalities close to 2, according to gel permeation chromatography (GPC) PS standards. Results in terms of conversion were excellent (above 70%). Oligomers were obtained with PDI around 1.8.

Similarly, Ishizu and Tahara [86] used allylmalonic acid diethylester as a transfer agent in the polymerization of MMA.

2.4.2
Oxidative Cleavage

Numerous authors [87, 88] have extensively studied the synthesis of telechelic oligomers by oxidative cleavage. Among them, Cheradame [89] reported the main reactions leading to telechelic polymers starting from high molecular weight. It was demonstrated that ozonolysis remains the most employed technique to get telechelic oligomers. In this field, lots of work has been done by Rimmer and Ebdon [88, 90–94]. For instance, they prepared telechelic oligo(2,3-dihydroxypropyl methacrylate acetonide) bearing aldehyde end groups by ozonolytic cleavage of poly(2,3-dihydroxypropan-1-methacrylate acetonide-*stat*-butadiene) followed by addition of methyl sulfide [94]. A typical MALDI-TOF spectrum of the expected aldehyde-telechelic oligomer was obtained. MALDI-TOF also revealed that no peak corresponded to oligomers containing a pendant aldehyde group that would be formed by ozonolytic cleavage of 1,2-butadiene units. However, MALDI-TOF proved the presence of a minor fraction of oligomers with α-acetaldehyde and ω-carboxylic acid end groups resulting from the use of dimethyl sul-

Table 12 Synthesis of α,ω-dihydroxy oligomers by ozonolysis cleavage of (co)polymers

Starting (co)polymer	Exp. conditions	Postozonitation treatment	Refs.
Poly(1-4 isoprene)	Inert solvent (– 70 to 30 °C)	Reduction with LiAlH$_4$	[370]
Acrylonitrile-butadiene copolymer	Tetrahydrofuran 15 °C	Reduction with Na borohydrides	[371]
Poly(isobutylene)	Cyclohexane	Reduction with H$_2$/Ni Raney	[372]
Poly(butene)	Suspension in hexane	Reduction with Ni Raney	[373, 374]
Poly(isobutylene)	Suspension in hexane	Thermal decomp. of peroxides	[375]

fide. These carboxylic acid oligomers were, however, completely removed by preparative ion exchange. Other activated copolymers, such as poly(MMA-*co*-butadiene) or poly(styrene-*co*-butadiene), were used to lead either to α,ω-dialdehyde PMMA or α,ω-dihydroxy PS [95–97].

The number of works in which authors took advantage of the reactivity of double bonds towards ozone to get telechelic oligomers is quite important. In Table 12, we have gathered some studies concerning the synthesis of α,ω-dihydroxy oligomers, presenting a peculiar interest in the research into new materials like polyurethanes.

2.5
Telechelic Oligomers Obtained by Atom Transfer Radical Polymerzation

LRP includes a group of radical polymerization techniques that have attracted much attention over the past decade for providing simple and robust routes to the synthesis of well-defined polymers, low-dispersity polymers, and the fabrication of novel functional materials [98–103]. The general principle of the methods reported so far relies on a reversible activation–deactivation process between dormant chains (or capped chains) and active chains (or propagating radicals). ATRP is a new method [8, 104, 105] allowing for the synthesis of telechelic oligomers [106]. The ATRP process can polymerize a wide range of monomers (styrene, acrylate, methacrylate, etc.). Furthermore, it allows the incorporation of reactive groups at the chain end of oligomers [107–110], such as amines [111], epoxides, or hydroxyl groups. ATRP is a radical process based on the use of a catalytic complex transition metal–ligand (Scheme 13). Generally the metal is copper (CuCl or CuBr), but Fe(II) [112] or Ru(II) [113, 114] may be used in some cases. The ligand (L) [114–116] is more often a tertiary amine. 1,1,4,7,10,10-Hexamethyltriethylenetetraamine (HMTETA) and 2,2′-bipyridine are commonly used [117], but Haddleton et al. [118, 119] also showed the good efficiency of *n*-(octyl)-2-pyridylmethanimine. The catalytic complex is able to establish equilibrium between the dormant species and the radicals. The equilibrium is shifted to the dormant species. The radical concentration is also low during the polymerization, which limits the termination reactions (disproportionation or recombination) and the number of dead chains.

In the first part of the reaction (Scheme 13), the catalytic complex will extract the halogen atom from the initiator (R–X) by a redox process. Then a radical is created that will propagate to the monomer. The growing chain will fix a halogen atom produced by the catalytic complex to form a dormant species. The chain will be reactivated when the catalytic complex traps the chain-end halogen atom to be oxidized. The chain is then able to propagate (Scheme 13). Some termination and transfer reactions occur but remain minor.

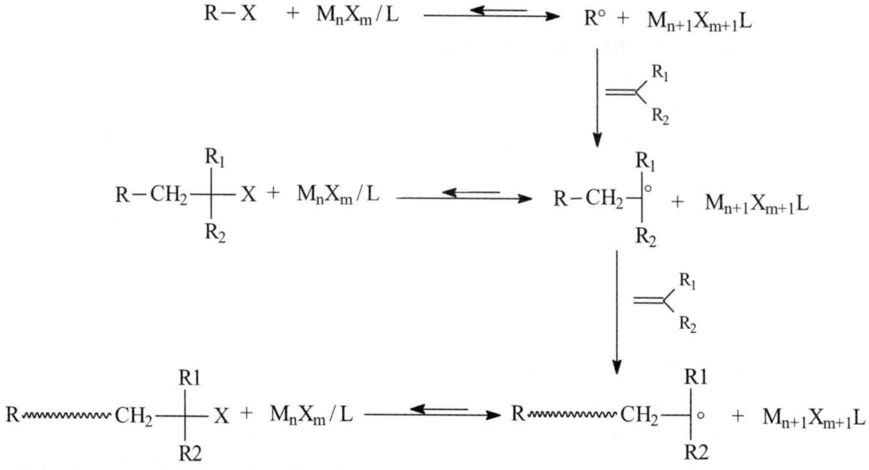

Telechelic Oligomers and Macromonomers by Radical Techniques

Scheme 13 Atom transfer radical polymerization (*ATRP*) mechanism

Getting the bifunctionality requires a chemical modification of the terminal halogen from the oligomer obtained by ATRP. Two different concepts are possible to obtain the bifunctionality (Scheme 14):

1. Chemical modification of the terminal halogen from a prepolymer (obtained by ATRP) bearing one halogen atom at a chain end and a reactive group at the other chain end
2. Chemical modification of the terminal halogen atoms from a prepolymer (obtained by ATRP) bearing one halogen atom at each chain end.

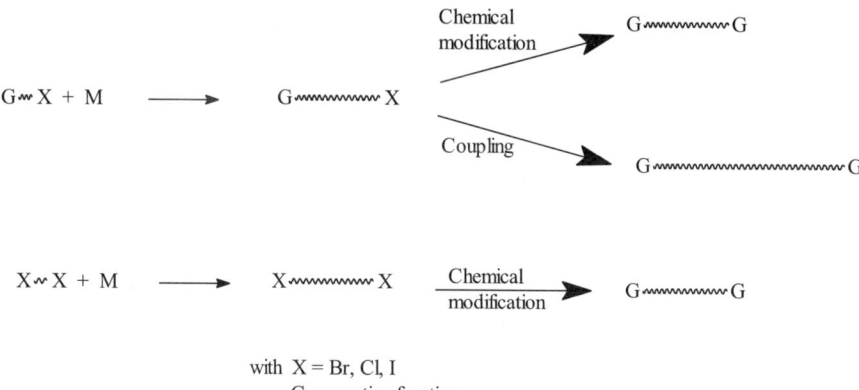

with X = Br, Cl, I
G = reactive function

Scheme 14 Synthesis of telechelic oligomers by chemical modification of prepolymers previously obtained by ATRP

2.5.1
Synthesis of Telechelic Oligomer Precursors

2.5.1.1
Synthesis of α-halogen Oligomers

The initiators used in ATRP have a similar structure, i.e., the halogen has to be in the β position of a carbonyl or aromatic group to make labile the C–X bond, with X being either a chlorine or a bromine atom. The control of this polymerization can be improved by the nature of the halogen, i.e., an initiator with a bromine atom exhibits a better reactivity than an initiator with a chlorine atom. These initiators can be used to obtain monofunctional oligomers. In the first case, the terminal oligomers will possess an R-group chain end provided by the initiator. This group can be an aldehyde [118], an amine [111, 118], a hydroxyl [120–122], a phenyl [118], a nitro [118], or an acid [123]. To get aliphatic acid and amine or anhydride functions, it is necessary to protect such groups [124–127]. Scheme 15 gives some examples of initiators used in ATRP.

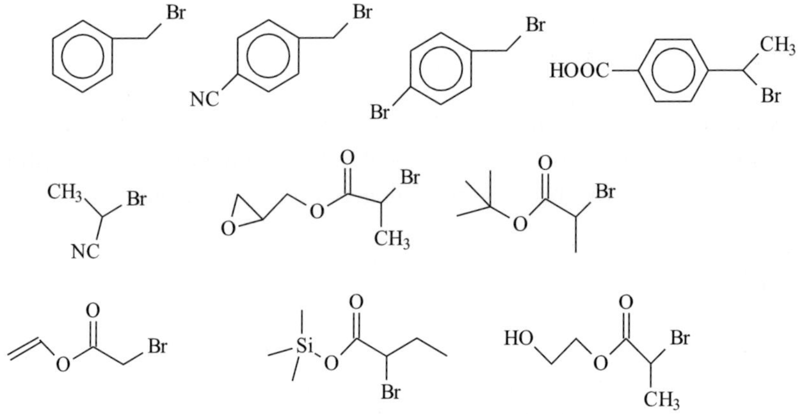

Scheme 15 Some examples of functional initiators used in ATRP

2.5.1.2
Synthesis of α,ω-dihalogen Oligomers

The synthesis of α,ω-dihalogen oligomers is realized by using bifunctional initiators, such as α,α-p-dihaloxylene [128] or arenesulfonyl chlorides [129]. Scheme 16 gives some examples of difunctional initiators.

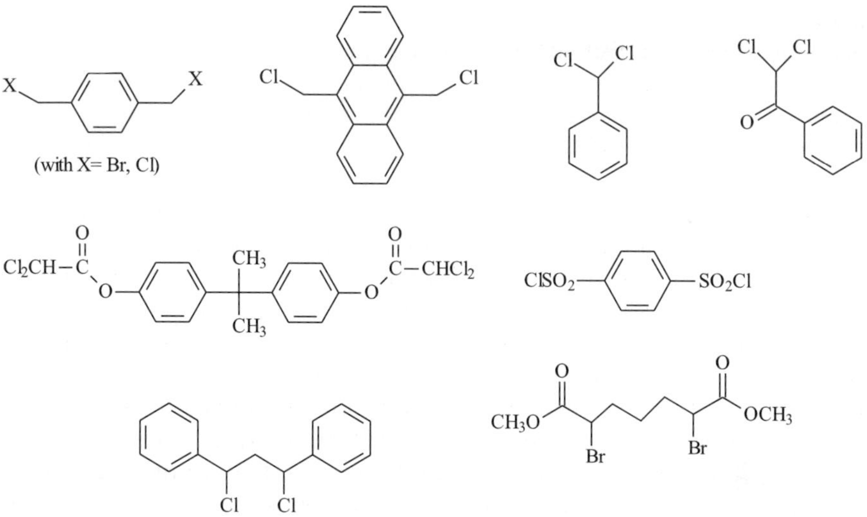

Scheme 16 Examples of difunctional initiators used in ATRP

2.5.2
Synthesis of Telechelic Oligomers

The chain-end halogen atom can be replaced by various methods, such as nucleophilic substitution or radical coupling. These techniques are developed next.

2.5.2.1
Nucleophilic Substitutions

The halogen end group can be transformed into other functionalities by means of standard organic procedures, such as a nucleophilic displacement reaction. Different authors have investigated this process of the nucleophilic displacement reactions with model compounds, to confirm the feasibility and selectivity. Compounds such as 1-phenylethyl halide, methyl 2-bromopropionate, and ethyl 2-bromoisobutane mimic the end groups of PSs, poly(alkyl acrylates), and poly(alkyl methacrylates), respectively. Different compounds have been tested, such as sodium azide, n-butylamine, and n-butylphosphine.

Azide End Groups
The reactions of the model compounds with sodium azide were performed in DMF at room temperature, with 1.1 equiv of sodium azide [130–132]. The kinetics of these reactions was followed by gas chromatography and the rate constants were calculated. The kinetics show that the reaction of the bromi-

nated substrates with sodium azide occurred almost instantly, but the chloro derivatives reacted about 100 times slower than the bromo derivatives; thus, the rates of the substitution reactions were dependent on the substrates. The primary carbon centers are the preferred sites of nucleophilic substitution reactions. But the reactivity of the secondary carbon centers may be enhanced by the electron withdrawing effect of ester groups. The authors used the collected data of the model studies to transform the chain-end halogen atom of polymers into reactive functions. These halogen end groups (chlorine and bromine) can be substituted by azide groups. Thus, PSs, poly(alkyl acrylates), and poly(alkyl methacrylates) with bromine end groups were reacted with sodium azide in solvents such as DMF or dimethyl sulfoxide (DMSO), which promoted nucleophilic substitution reactions. A complete substitution of the bromine by azide was observed by MALDI-TOF and ^{1}H NMR analysis. In the case of poly(alkyl methacrylates) an excess of sodium azide was necessary.

This azide group can be reduced with lithium aluminum hydride and converted into amine end groups (Scheme 17); however, this procedure could not

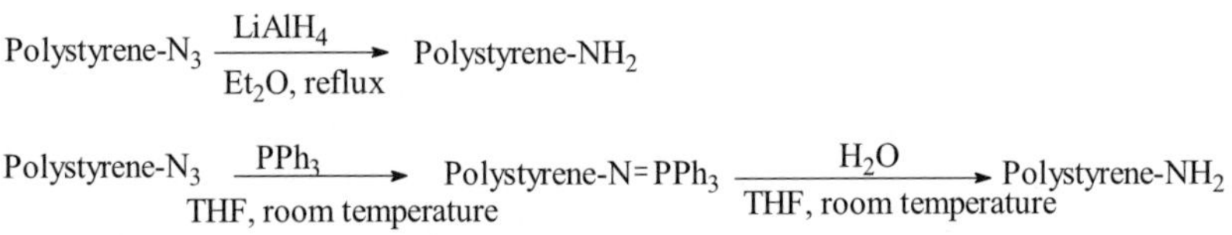

Scheme 17 Reduction of azide group

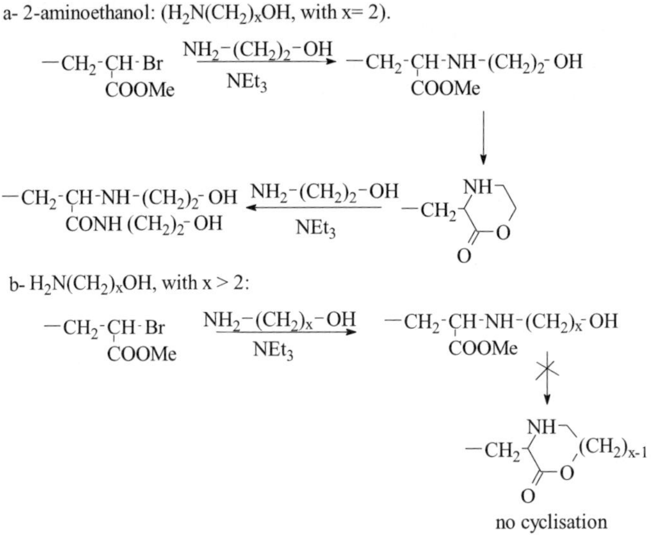

Scheme 18 Functionalization of poly(methyl acrylate)-Br with two agents: 2-aminoethanol ($x = 2$) and 5-amino-1-pentanol ($x = 5$)

be used for poly(alkyl acrylates) and poly(alkyl methacrylates), because the reduction of the ester functionalities may occur.

An other method, described by Coessens et al. [130], is the conversion of the azide group into the phosphoranimine end groups and subsequent hydrolysis to the amino end groups (Scheme 18). This procedure was used to synthesize diamine telechelic oligomers of PSs. Styrene was initiated by a difunctional initiator (α,α'-dibromo-p-xylene) yielding α,ω-dibromo PSs. The bromine atoms are then converted into amino end-groups [123].

It is also possible to use the trimethylsilyl azide in the presence of tetrabutylammonium fluoride to transform the terminal halide into an azide group.

Amino End Groups

As shown in the previous section, the halogen end groups of polymers prepared by ATRP can be substituted by good nucleophiles such as azides. But, Coessens and Matyjaszewski [133] showed that the direct displacement of a halogen by a hydroxide anion is followed by side reactions such as elimination. However, the authors described the nucleophilic substitution of the halogen end group by the primary amine, i.e., 2-aminoethanol to introduce other functionalities. The primary amine gives good and selective nucleophiles to substitute the bromine end groups of PS oligomers [133, 134], but these reactions were tested with poly(alkyl acrylates) and poly(alkyl methacrylates) at room temperature in DMSO. The authors demonstrated that the substitution reaction altered the ester function. Thus, a selective substitution of the bromine end groups of PS by 2-aminoethanol was expected.

The reaction of poly(methyl acrylate)-Br with 2-aminoethanol was expected to result in multiple substituted products. This result was ascribed to the fact that after the substitution of the bromine by 2-aminoethanol, formation of a six-membered ring could occur (Scheme 18). Afterwards, ring opening by attack of a second 2-aminoethanol molecule could lead to the double-substituted product. The α-bromo poly(methyl acrylate) could be suppressed by using 4-aminobutanol instead of 2-aminoethanol as a nucleophile, without side reactions (Scheme 18). For example, the yield of functionalization of poly(n-butyl acrylate) with 5-amino-1-pentanol is close to 96% [197].

Thiol End Groups

The halogen functional polymer can react with a thiol by nucleophilic reaction, resulting in a polymeric thioether and a hydrogen halide. The latter is trapped by a basic additive, preventing a reverse reaction. Snijder et al. [135] used this technique to modify the end group of poly(n-butyl acrylate) into a hydroxy-functional polymer. With 2-mercaptoethanol, the yield of functionalization was higher with the addition of 1,4-diazabicyclo[2,2,2]octane (DABCO) to the reaction mixture. The addition of DABCO allows for the formation of a sulfide anion, which is a stronger nucleophile. They studied this

Table 13 Functionalization of bromo poly(n-butyl acrylate) (*PBA-Br*) using nucleophilic substitution in the presence two functional agents: 2-mercaptoethanol and 5-amino-1-pentanol

Functional agents	Experimental conditions (mol l^{-1})			Functionality
	[PBA-Br]	[Functional agent]	[DABCO]	
HO-(CH$_2$)$_2$-SH	1×10^{-2}	2×10^{-2}	–	0.14
HO-(CH$_2$)$_2$-SH	1×10^{-2}	2×10^{-2}	2×10^{-2}	0.96
HO-(CH$_2$)$_5$-NH$_2$	3.8×10^{-2}	5.5×10^{-1}	–	0.96

DABCO 1,4-diazabicyclo[2,2,2]octane

modification and this mechanism by gradient polymer elution chromatography. The rate constants of the functionalization reaction were determined by this last technique. The values of the rate constant and the functionality are given in Table 13.

By comparing different methods, Snijder et al. [135] showed that the coupling afforded the best results.

2.5.2.2
Radical Addition Reactions

Allyl Tri-*n*-butylstannane

A one-pot process to displace the halogen end groups by allyl end groups was developed using allyl tri-*n*-butyltin. The reaction of an alkyl halide with allyl tri-*n*-butyltin is a radical reaction that tolerates the presence of other functional groups such as acetals, ethers, epoxides, and hydroxyl groups. This technique was also used for the deshalogenation of polymers prepared by ATRP (Scheme 19).

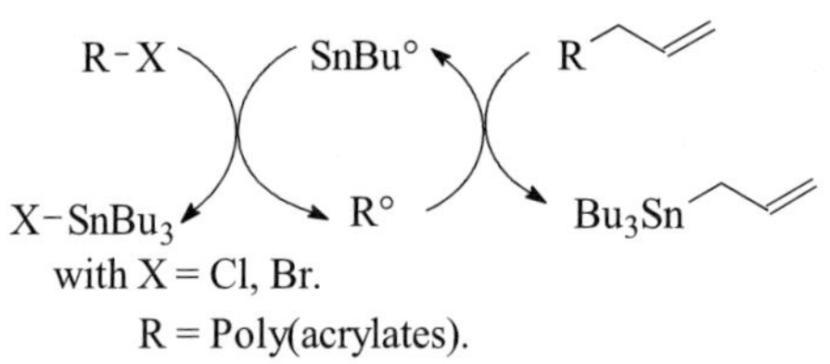

$$R-X \quad SnBu° \quad R\diagup\!\!\!\diagdown$$
$$X-SnBu_3 \quad R° \quad Bu_3Sn\diagup\!\!\!\diagdown$$

with X = Cl, Br.
R = Poly(acrylates).

Scheme 19 Reaction of allyl tri-*n*-butylstannane with alkyl halides [348]

For example, poly(alkyl acrylates) with bromine end groups were reacted with allyl tri-*n*-butyltin and Cu(0) in benzene. After 3 h, complete radical addition reaction was obtained. ^{1}H NMR confirmed the presence of the allyl function.

Incorporation of Less Reactive Monomers

1,2-Epoxy-5-hexene and allyl alcohol [133] are some examples of monomers not polymerizable by ATRP. The main reason is that with the catalytic systems used in ATRP, the activation process is too slow because the radical formed is not stabilized by resonance or by electronic effects. However, when these monomers were added at the end of the polymerization reaction of acrylates [133, 136] or methacrylates [128], the radicals of the poly(alkyl acrylate) chain end were able to add to these monomers and the deactivation provided halogen-terminated polymers. These radical addition reactions can occur owing to the rate constants of poly(methyl acrylate). This polymer was previously obtained by ATRP with 95% conversion, using an excess of 1,2-epoxy-5-hexene (25-fold excess towards the end groups). At the same time, Cu(0) (0.5 equiv towards CuBr) was added in order to reduce the amount of Cu(II) in the reaction mixture. Less Cu(II) in the reaction mixture results in a faster radical reaction; however, too high Cu(I) or too low Cu(II) concentrations can result in bimolecular termination reactions and incomplete functionalization. After the reaction, the polymer was purified by filtering through alumina. Electrospray ionization mass spectroscopy (MS) demonstrated that the epoxide was incorporated at the chain end.

Similar reaction conditions were used with allyl alchohol [131, 133, 136, 137], and the addition of allyl alcohol to the poly(alkyl acrylate) chain is shown in Scheme 20 [137].

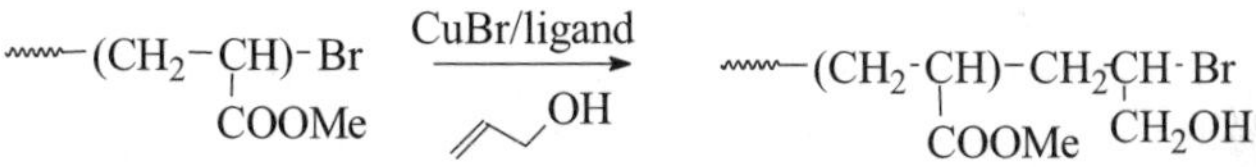

Scheme 20 Addition of allyl alcohol to polyacrylates

Other less reactive monomers were incorporated into chain ends of oligomers, including divinyl benzene for MMA [138] and maleic anhydride for styrene [139] (Scheme 21) and methacrylates [140]. The method for the synthesis of maleic anhydride terminated PS is based on the fact that maleic anhydride cannot be homopolymerized under normal conditions [141]. These maleic anhydride terminated PSs were used for the compatibilization of the nylon 6–PS binary system in the melt by reaction with NH_2-terminated polyamide.

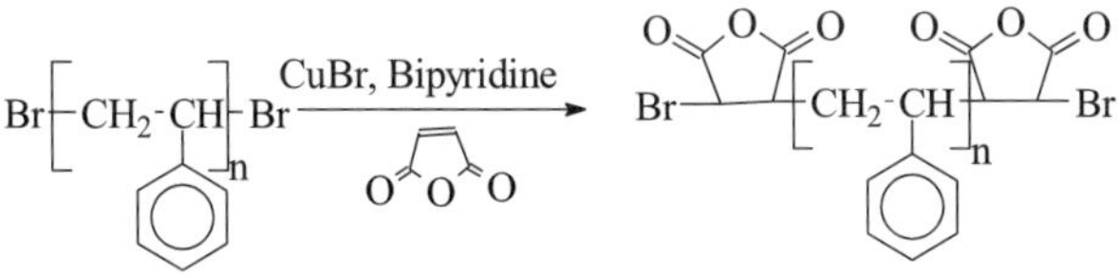

Scheme 21 Addition of maleic anhydride to polystyrene

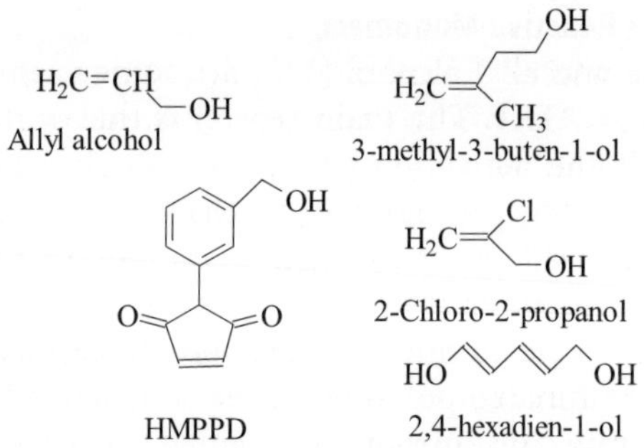

Scheme 22 Examples of less reactive monomers used to give the functional oligomer

Other radicals could be used to give the functional oligomer, such as 2-chloro-2-propenol, 2,4-hexadien-1-ol [142], and 3-methyl-3-buten-1-ol [143] (yield of reaction is 20%) (Scheme 22).

Functionalization by "Click" Chemistry

Recently, the group of Sharpless [144, 145] popularized the 1,3-dipolar cycloaddition of azides and terminal alkynes, catalyzed by copper(I) in organic synthesis. This process was proven to be very practical, because it can be performed in several solvents (polar, nonpolar, protic, etc.) and in the presence of different functions. These cycloadditions were classified as "click" reactions, defined by Sharpless.

The click chemistry is a very practical process for the synthesis of new polymers [146] or postfunctionalized polymers [147–149]. It allows for the synthesis of telechelic compounds by transformation of the halogen end

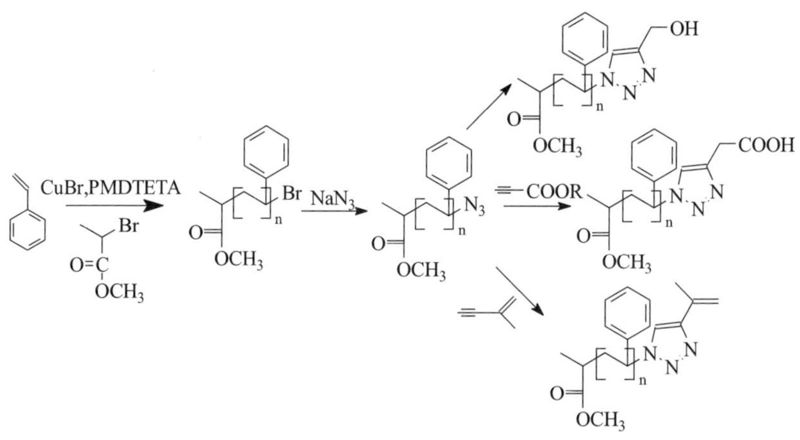

Scheme 23 Transformation of bromine end functional polystyrene into various functional groups by "click" chemistry

group that is easily transformed into an azide function. In a second step, the 1,3-cycloaddition of end-functional azide polymers to functional alkynes is a versatile method for the preparation of various end-functional polymers (Scheme 23). Lutz et al. [150, 151] used this technique for functionalizing oligomers of PS ($M_n = 2700\,\text{g mol}^{-1}$). The synthesis is performed in tetrahydrofuran (THF) in the presence of a CuBr and 4,4'-di-(5-nonyl)-2,2'bipyridine (dNbipy) complex. The choice of this ligand is very important, because it could accelerate the catalysis of cycloaddition [152]. This technique can therefore be considered as a "universal" method and allows for a quantitative transformation of the PS chain end into the desired function [150, 151].

2.5.2.3
Use of a Quencher Agent

Addition of Excess Initiator at the End of the Polymerization
A one-pot synthesis of telechelic and semitelechelic poly(alkyl acrylates) with unsaturated end groups has been developed by Bielawski et al. [1]. ATRP of methyl acrylate or n-BA was initiated with either ethyl α-bromomethylacrylate or methyl dichloroacetate, as a monofunctional or a difunctional initiator, respectively, and was mediated with various Cu–amine complexes. Addition of excess ethyl 2-bromomethylacrylate was found to immediately quench the polymerization, but also to insert 2-carbethoxyallyl moieties at the ends of the polymer chains (Scheme 24). Thus, the synthesis of telechelic poly(alkyl acrylates) with unsaturated end groups has been accomplished, with very good functionality ($f \approx 2$) (Scheme 24).

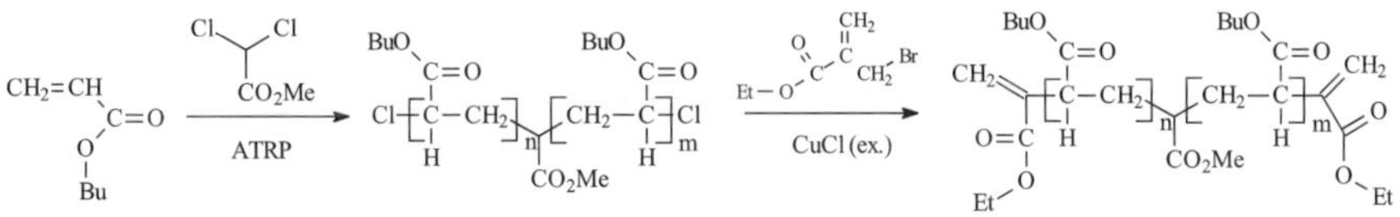

Scheme 24 ATRP of n-butyl acrylate initiated with a commercially available difunctional initiator and chemical modification of the chain end by addition of ethyl 2-bromomethylacrylate excess

Silyl Enol Ether
Sawamoto's group used this process for the first time in 1998. Silyl enol ethers such as p-methoxy-α-(trimethylsilyloxy)styrenes [153] or isopropenoxytrimethylsilane [153] are efficient quenchers in the LRP of MMA using the $RuCl_2(PPh_3)_3$ complex. They convert the C – X, X being a halogen atom, into a C – C bond with a ketone group. As shown in Scheme 25, the silyl compound mediated quenching reaction probably proceeds via the addition of the growing radical into the C = C double bond of the quencher, followed by

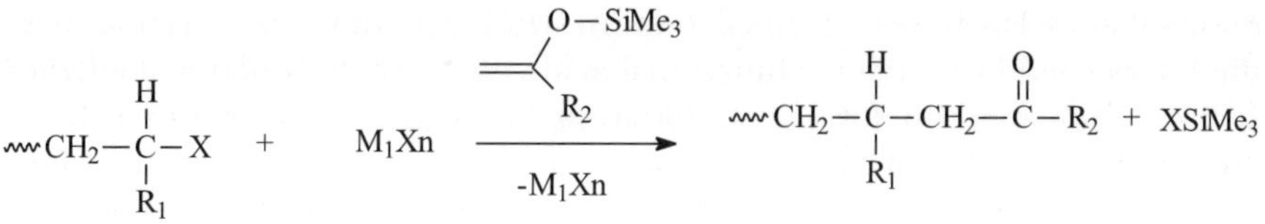

$$\text{\textasciitilde} CH_2-\underset{R_1}{\underset{|}{\overset{H}{\overset{|}{C}}}}-X \quad + \quad M_1Xn \quad \xrightarrow[-M_1Xn]{\underset{R_2}{\overset{O-SiMe_3}{=\!\!\!<}}} \quad \text{\textasciitilde} CH_2-\underset{R_1}{\underset{|}{\overset{H}{\overset{|}{C}}}}-CH_2-\overset{O}{\overset{||}{C}}-R_2 \quad + \quad XSiMe_3$$

with $R_1 = CO_2CH_3$ and R_2 = Ph, Ph.OCH$_3$, Ph.Cl, Ph.Br.

Scheme 25 Reaction of end capping (silyl enol ether) agents onto poly(alkyl methacrylates)

the elimination of the silyl group and the chlorine that originated from the terminal polymer.

Ando et al. [108, 154] suggested this method for the functionalization of a PMMA oligomer obtained by transition-metal-mediated LRP. MMA was polymerized with a binary initiating system consisting of dimethyl 2-chloro-2,4,4-trimethylglutarate initiator and $RuCl_2(PPh_3)_3$ in the presence of aluminum triisopropoxide in toluene at 80 °C. After this polymerization, the quenching reaction is considered to proceed from the growing radical to the vinyl group to generate another terminal radical, followed by elimination of a trimethylsilyl group with the chlorine at the polymer chain end, owing to its high affinity toward halogens.

In the case of p-substituted-α-(trimethylsilyloxy)styrenes [153], the quenching is selective and quantitative. Thus, the quenching proceeds faster with an electron-donating susbtituent (OCH_3 > H > F > Cl) on phenyl groups. The reaction is favored with these silyl enol ethers by the presence of the α-phenyl group, which stabilizes the radical by the electron-donating effect of the aromatic group after the addition of the quencher double bond. Indeed, the phenyl group increases the electron density and the reactivity of its double bond.

In contrast, silyl enol ethers with an R-alkyl group (R-silyloxy vinyl ethers) proved to be less efficient, indicating that the stability of the resultant silyloxyl radical is the critical factor for the design of good quenchers. This is due to the degree of affinity of the PMMA radical towards the vinyl groups in the quenchers. This silyl enol ether capping is applicable for copper-catalyzed polymerizations, carried out on isolated PMMA. The quenching has been carried out not in situ but on isolated PMMA samples. The trimethylsilyloxy group at the 4-position can also be converted into the phenol function. An interesting application of the silyl enolate capping reaction has been developed by Percec [120, 155], who coined the "TERMINI" capping agents (irreversible termination multifunctional initiator). This refers to a "protected functional compound able to quantitatively terminate a living polymerization and, after deprotection, to quantitatively reinitiate the same or a different living polymerization in more than one direction."

Addition of Stable Radicals at the End of the Polymerization

Stable radicals, such as nitroxides:hydroxy-2,2,6,6-tetramethylpiperidinyloxy (TEMPO) [8, 156], can be added to the polymerization medium to terminate all polymer radicals produced. For styrenes and acrylates [157], this mainly occurs through combination. Chambard et al. [157] showed this technique allows for the modification of poly(n-butyl acrylate)-Br in the presence of an excess of hydroxy-TEMPO, resulting in hydroxy-functional poly(n-butyl acrylate) with good functionality ($f > 95\%$). This process is not desirable, because the polymer produced is thermally unstable (carbon nitroxide) and cannot be used at high temperature.

2.5.2.4 Coupling Reactions

Radical Coupling

This reaction is based on the Wurtz [158, 159] radical coupling. Two teams developed at the same time this radical coupling, also called atom transfer radical coupling (ATRC) [160–164].

This reaction takes place in the presence of a transition metal such as copper or iron and consists of coupling α-halogen oligomers, previously synthesized by an ATRP process (Scheme 26).

This reaction was first performed on molecules suitable for modeling the chain end of oligomers. For instance, Otazaghine et al. [160, 161] performed the radical coupling of 1-bromoethylbenzene at 65 °C in anisole, in the presence of Cu(0) and HMTETA, with a quantitative yield. They then applied the same experimental conditions to α-halogen oligomers of PS. The coupling yield was almost 100%, confirmed by the disappearance of the CH – Br signal in ^{1}H NMR. They also observed a doubling of the molecular weight by GPC (Fig. 2). MALDI-TOF analysis confirmed the expected telechelic structure.

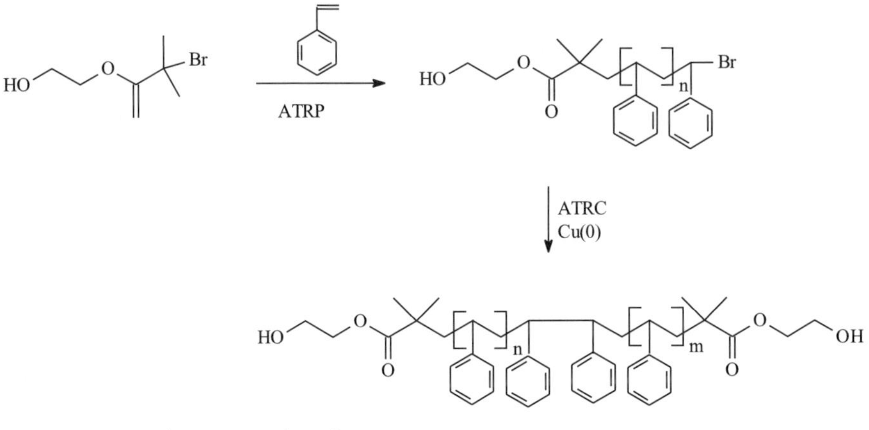

Scheme 26 Coupling process based on ATRC

In a similar way, α-halogen oligomers of acrylates [162], previously synthesized by ATRP, were used in ATRC; however, the yield of the radical coupling was lower than that of styrene (Table 14). A similar result was observed for

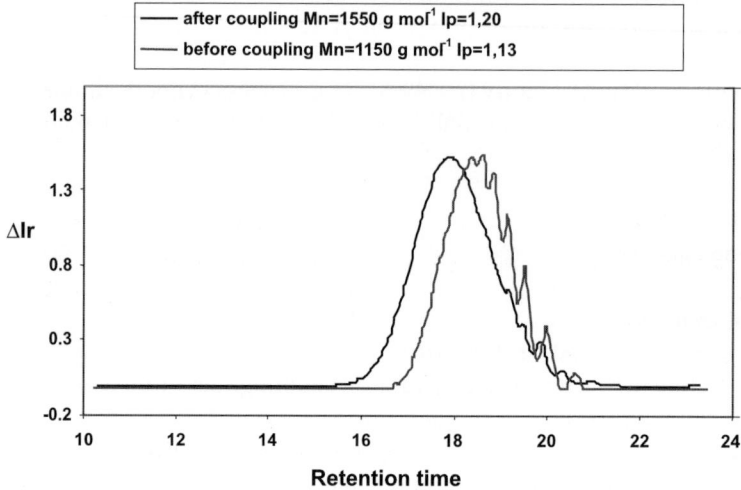

Fig. 2 Gel permeation chromatography of monomobrominated oligomers of polystyrene using 2SbiB as the initiator and of the products from the coupling reaction

Table 14 Percentage of coupled chain for different oligomers

Oligomers	M_n (g mol^{-1})	Experimental conditions (mol l^{-1})			Coupling yield (%)
		[Cu(0)]	[CuBr]	[Ligand]	
Polystyrene	1550	1	0	2[b]	70 [160, 161]
(PStBr)	1780	4	1	1[a]	87 [163]
	1780	4	1	2[a]	94 [163]
	1780	4	1	5[a]	99 [163]
Poly(methyl acrylate)	950	1	0	2[b]	67 [162]
	1280	1	0	2[b]	67 [162]
Poly(n-butyl α-fluoro-acrylate)	1550	1	0	2[b]	78 [160, 161]
Poly(n-butyl acrylate)	900	2	0	2[b]	63 [162]
	1150	2	0	2[b]	62 [162]
	1850	2	0	2[b]	59 [162]
Poly(methyl methacrylate)	2000	2	0	2[a]	
	2000	2	0	2[b]	

[a] Ligand PMDETA
[b] Bipyridine

Table 15 Percentage of coupled chain for different number-average molar masses of n-butyl acrylate terminated by styrene

Oligomers	M_n (g mol^{-1})	Experimental conditions (mol l^{-1})			Coupling yield (%)
		[Cu(0)]	[CuBr]	[Ligand]	
Poly(n-butyl acrylate-b-styrene)-Br	1510	1	0	2	75.10
	1820	2	0	2	75.7
	2350	2	0	2	73.9

α-fluoroacrylate monomers [160, 161]. Concerning the ATRC of α-halogen oligomethacrylates, the coupling does not occur, owing to steric effects [162].

To increase the coupling rate, some styrene units were incorporated at the chain end of α-halogen oligoacrylates. The coupling yield then approached 100% (Table 15).

Coupling Through Nucleophilic Substitution

Yurteri et al. [164] suggested a different approach. The coupling of the oligomers is realized by using organic molecules such as hydroquinone in the presence of K_2CO_3 and DMF. They showed that a quantitative coupling rate required an exact stoichiometry (Scheme 27).

In conclusion of this part, ATRP is a new versatile method, leading to the synthesis of precursors for telechelic oligomers. The chain-end halogen atom is chemically modified to obtain the telechelic structure; hence, getting the telechelic structure also requires the accuracy of the halogen functionality. Only a few studies were concerned with following the dependence of the functionality upon the reaction time. Lutz et al. [150, 151] measured the chain-end bromine functionality by ^{1}H NMR for the ATRP of styrene in the presence of dNbipy. The oligomers obtained exhibited a molecular weight of about 10 000. They observed a linear decrease of the functionality upon the monomer conversion. Moreover, for conversions up to 90%, the functional-

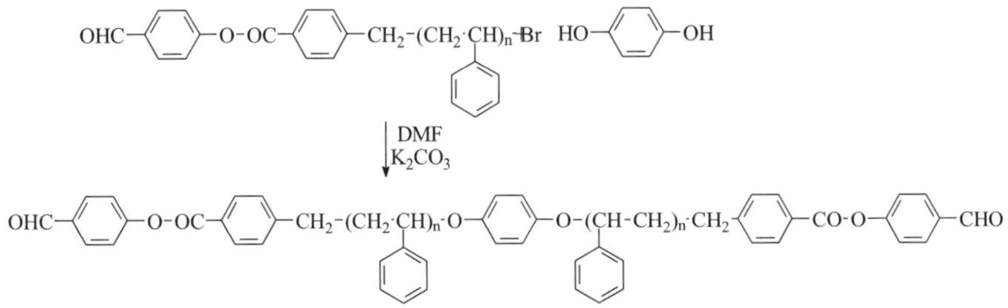

Scheme 27 Coupling of α-bromo polystyrene oligomers by nuclepophilic substitution using hydroquinone in the presence of K_2CO_3 in dimethylformamide

ity dramatically dropped. The authors confirmed these results by a simulation (using PREDICI software).

Different reactions may affect the chain-end bromine atom of PS during ATRP: transfer process, bimolecular terminations, or elimination reactions induced by the Cu(II) complex. The authors showed that the loss in functionality was predominantly due to β-hydrogen elimination reactions. This result is very important for the synthesis of telechelic polymers by ATRP, because all processes (described later) are based on the halogen transformation.

2.6
Telechelic Oligomers
Obtained by Reversible Addition–Fragmentation Chain Transfer

Among the LRP, RAFT [13, 165–167], and macromolecular design by interchange of xanthates (MADIX) [168] homologues concerning the xanthate species are versatile techniques to produce polymer architectures, such as telechelic ones. RAFT and MADIX are both based on a radically induced degenerative transfer reaction, first reported by Zard's group [169], between a thiocarbonyl-thio containing compound and a propagating radical. The mechanism of RAFT, proposed by Chiefari et al. [170], consists of many complex equilibrium steps and involves a rapid exchange of the radical among all the growing polymeric chains via addition–fragmentation reaction with the CTA. The mechanism of the RAFT process, presented in Scheme 28, is very

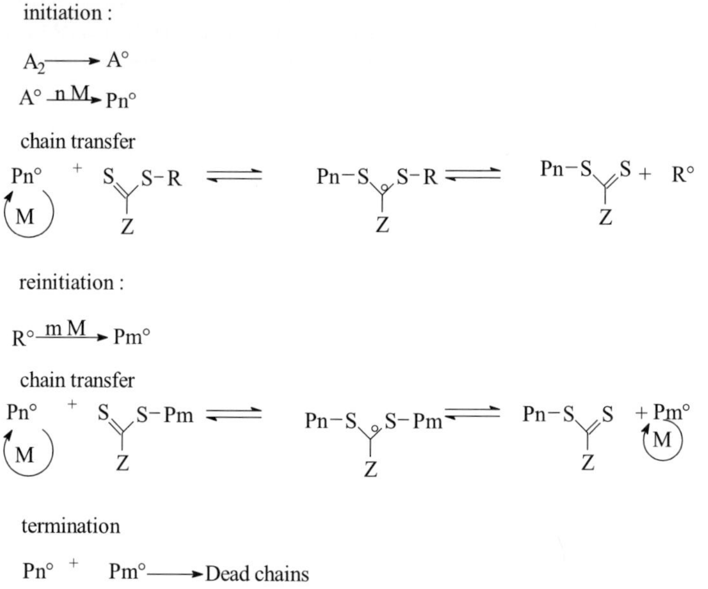

initiation :

$A_2 \longrightarrow A°$
$A° \xrightarrow{nM} Pn°$

chain transfer

$Pn° + S\!\!\!=\!\!\!C(Z)\!-\!S\!-\!R \rightleftharpoons Pn\!-\!S\!-\!C(Z)\!-\!S\!-\!R \rightleftharpoons Pn\!-\!S\!-\!C(Z)\!=\!S + R°$
(M)

reinitiation :

$R° \xrightarrow{mM} Pm°$

chain transfer

$Pn° + S\!\!\!=\!\!\!C(Z)\!-\!S\!-\!Pm \rightleftharpoons Pn\!-\!S\!-\!C(Z)\!-\!S\!-\!Pm \rightleftharpoons Pn\!-\!S\!-\!C(Z)\!=\!S + Pm°$
(M)   (M)

termination

$Pn° + Pm° \longrightarrow$ Dead chains

Scheme 28 Reversible addition–fragmentation chain transfer polymerization (*RAFT*) or macromolecular design by interchange of xanthates mechanism complex and was recently investigated in detail (in terms of kinetics parameters) by several authors [171, 172].

RAFT, allowing for predictable molecular weight with low polydispersities, is applicable to a wide range of vinyl monomers [173–176], some of them not always being polymerizable by NMP or ATRP (i.e., VAc [167] or monomers bearing protonated acid groups). Hence, RAFT is employed in many polymerization processes, such as bulk, solution, suspension, emulsion, and miniemulsion [177–180].

For achieving the chain-end functionality in the polymer by RAFT, it is necessary either to adjust the structure of the transfer agent or to combine it with a modification of the terminal dithioester. Scheme 29 summarizes two different pathways for getting the telechelic structure. In the first pathway, the transfer agent is a trithioester compound, bearing two leaving groups R^1. The telechelic structure is directly obtained with an expected trithioester group at the middle of the polymer. The second pathway considers a dithioester as the transfer agent, bearing a leaving group R^1 at one end and a nonliving group at the other end. After RAFT onto the monomer M_1, the polymer contains the chain-end R^1 group but also the chain-end thioester. The bifunctionality is then obtained by chemically modifying the chain-end thioester into a chain-end R^1 group.

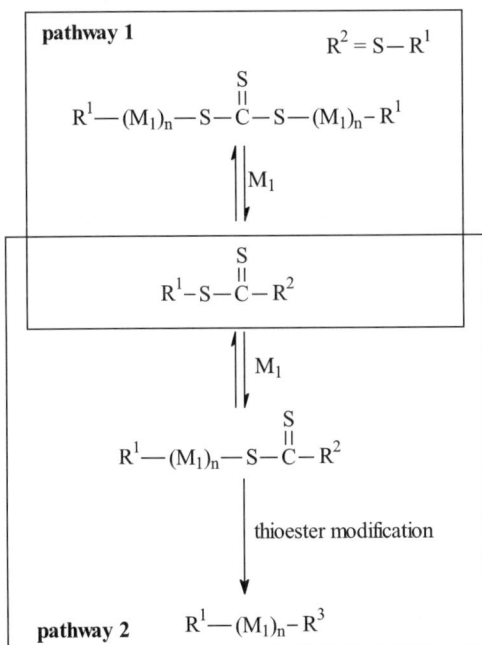

Scheme 29 Synthesis of telechelic polymers by the RAFT mechanism (R^1 and R^3 being functional groups)

We will investigate the two different pathways (Scheme 29) leading to a telechelic structure by RAFT.

2.6.1
Use of a Trithioester Transfer Agent

Although only one step is necessary to get to the desired telechelic structure, this pathway is much less developed than the chemical modification of the terminal dithioester function. This may be attributed to a noneasy synthesis of such trithioester transfer agents [181, 182] and also to the bad stability of this group located at the center of the molecule. Hence, the first trithioesters used in the RAFT process were aimed at getting multiblock copolymers [183, 184]. The R^1 leaving group was not generally a suitable function for further polycondensation reactions. It is also noteworthy that trithioesters can be employed in an aqueous medium [185, 186]. As an example, Baussard et al. [184] synthesized a new trithioester, sodium S-benzyl-S'-2-sulfonatoethyl trithiocarbonate. This trithiocarbonate was employed as transfer agent for the RAFT of vinylbenzyltrimethylammonium chloride in an aqueous medium. Although good control of the polymerization occurred, the benzyl end group is not suitable for polyaddition reaction. Some trithioester compounds, however, have suitable end groups for further polycondensation reactions of the telechelic oligomers obtained. J. Liu et al. [187] and R.C.W Liu et al. [188] synthesized a novel RAFT agent, S,S'-bis(2-hydroxyethyl-2'-butyrate)trithiocarbonate (BHEBT), bearing two hydroxyl end groups. The synthesis is rather complex and is done through a three-step reaction involving the presence of an anion-exchange resin with OH^- (Scheme 30).

P. Liu et al. [27], Z. Liu et al. [94], J. Liu et al. [187], and R.C.W. Liu et al. [188] performed the synthesis of hydroxy-telechelic PS or poly(methyl acrylate) by direct RAFT of styrene and methyl acrylate, respectively, with BHEBT. BHEBT was proven to be a highly efficient transfer agent towards styrene and methyl acrylate by plotting M_n against monomer conversion. PDIs were found to be less than 1.2. The authors demonstrated that the trithiocarbonate group was in the middle of the polymer chain because of the similar fragmentation reactivity of the two leaving groups, 2-hydroxylethyl-2'-butyrate. Finally the telechelic structure was proved for both styrene and methyl acrylate by means of ^{1}H NMR.

As another example, Lai et al. [189] reported the synthesis of carboxyl-terminated trithiocarbonates. The synthesis of S,S'-bis($\alpha\alpha'$-dimethylacetic acid)trithiocarbonate is presented in Scheme 31 and requires the use of carbon disulfide, which reacts with hydroxide ions. This synthesis yields trithiocarbonate with purities above 99%.

This trithiocarbonate is expected to exhibit high chain transfer efficiency and good control over the RAFT because the living group corres-

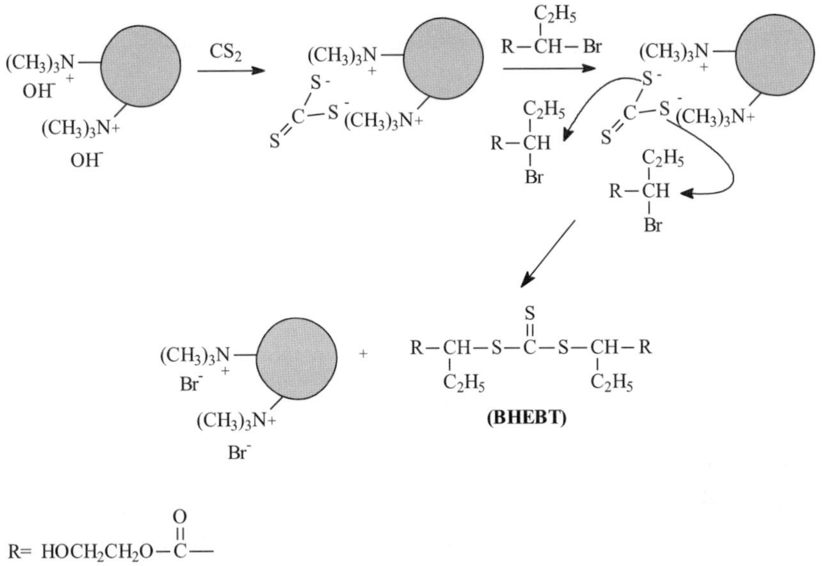

Scheme 30 Synthesis of S,S'-bis(2-hydroxyethyl-2'-butyrate)trithiocarbonate

Scheme 31 Synthesis of carboxyl-terminated trithiocarbonate

ponds to a tertiary carbon and bears a radical-stabilizing carboxyl group. Lima et al. [190–192] polymerized butyl acrylate by using this carboxyl-terminated trithiocarbonate. Quantitative yields were obtained with M_n around 2000 g mol^{-1} for PDI less than 1.15, depending on the experimental conditions. To determine the fraction of telechelic poly(n-butyl acrylate) Lima et al. [192] developed a new liquid chromatography (LC) [193–195] method, based on the carboxyl end-group functionality (the retention being independent of M_n). LC separations revealed the dominant presence of bifunctionality (more than 97%). The low number of monofunctional chains was due to side reactions inherent to growing radicals, such as bimolecular recombination or disproportionation [196]. The complete bifunctionality was obtained when ACVA was used as the initiator.

More recently, Jiang et al. [197, 198] used critical LC coupled with MS to determine three main structures of carboxyl-terminated PBA, depending on the initiator (Scheme 32).

Finally, Convertine et al. [199] illustrated the use of S,S'-bis($\alpha\alpha'$-dimethylacetic acid)trithiocarbonate for the RAFT of both acrylamide and N,N-

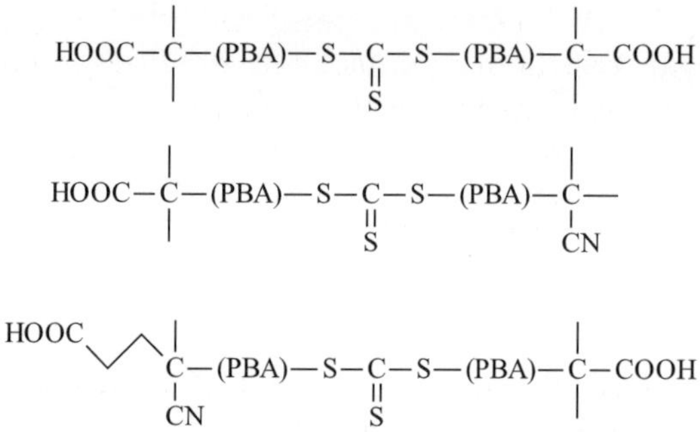

$$\text{HOOC}-\underset{|}{\overset{|}{\text{C}}}-(\text{PBA})-\text{S}-\underset{\text{S}}{\overset{\|}{\text{C}}}-\text{S}-(\text{PBA})-\underset{|}{\overset{|}{\text{C}}}-\text{COOH}$$

$$\text{HOOC}-\underset{|}{\overset{|}{\text{C}}}-(\text{PBA})-\text{S}-\underset{\text{S}}{\overset{\|}{\text{C}}}-\text{S}-(\text{PBA})-\underset{\text{CN}}{\overset{|}{\text{C}}}-$$

$$\underset{\text{CN}}{\overset{\text{HOOC}}{\diagup\!\!\diagdown}}\underset{|}{\overset{|}{\text{C}}}-(\text{PBA})-\text{S}-\underset{\text{S}}{\overset{\|}{\text{C}}}-\text{S}-(\text{PBA})-\underset{|}{\overset{|}{\text{C}}}-\text{COOH}$$

Scheme 32 Main structures of carboxy-functional poly(n-butyl acrylate) (*PBA*) synthesized by RAFT polymerization initiated with either α,α'-azobis(isobutyronitrile) or 4,4′-azobis(4-cyanovaleric acid)

dimethylacrylamide in aqueous media at room temperature. They showed that RAFT was conducted to high conversion with a living character. The dicarboxyl functionality was evidenced.

This first pathway using a trithioester transfer agent afforded functionality close to 2. However, the final oligomer contains a trithioester group in the middle of the chain that is highly labile. A further polycondensation, which requires high temperature, with such oligomers, obtained by this technique, seems not to be favored.

2.6.2
Thioester Modification

The second pathway combined at least a two-step reaction: the first step is the RAFT in the presence of a dithioester transfer agent, whereas the second step consists of the removal of the thioester terminal (Scheme 29). The first step occurs with RAFT agents bearing only one polycondensable function. Such transfer agents are numerous [200, 201] compared with their trithioester homologues, even if the syntheses are usually costly and require multistep reactions. Table 16 gives an overview of dithioester compounds found in the literature and leading to a monofunctional oligomer by RAFT.

Among all these RAFT reactions, only a few workers have been interested in removing the thioester end group; hence, monofunctional oligomers presented in Table 16 are of interest because they become macrothioester transfer agents and offer the possibility of synthesizing diblock copolymers through other RAFT processes. The removal of the terminal thioester group is more complicated than undertaking a RAFT because it involves chemical modification followed by a purification of the new difunctional oligomer.

We will show a few examples illustrating the synthesis of telechelic oligomers by modification of the terminal thioester group.

Lima et al. [190–192] performed the RAFT of MMA in the presence of (4-cyano-1-hydroxylpent-4-yl) dithiobenzoate CTA (entry E in Table 16). Monohydroxy oligomethylmethacrylates were obtained. To get the telechelic structure, aminolysis of monofunctional PMMA with 1-hexylamine was undertaken, leading to a α-OH,ω-SH-PMMA. A hydroxyl group can replace the thiol terminal by Michael addition with hydroxyethyl acrylate [192] (Scheme 33).

Table 16 Some dithioester CTAs used in reversible addition–fragmentation chain transfer polymerization to give monofunctional oligomers

Entry	CTAs	Monomer	Polymer characteristics	Refs.
A	(dithiobenzoate with C(CH$_3$)$_2$CN)	Styrene	$M_n = 20 \times 10^3$ g mol^{-1} PDI = 1.18	[202]
		Acrylamide	$M_n = 36 \times 10^3$ g mol^{-1} PDI = 1.23	
B	(dithiobenzoate with CH(Ph)C(O)OMe)	Styrene	$M_n = 13 \times 10^3$ g mol^{-1} PDI = 1.10	[202]
		MA	$M_n = 13 \times 10^3$ g mol^{-1} PDI = 1.18	
C	(dithiobenzoate with CH(Ph)C(O)NEt$_2$)	MA	$M_n = 48 \times 10^3$ g mol^{-1} PDI = 1.21	[202]
		MA/styrene	$M_n = 42 \times 10^3$ g mol^{-1} PDI = 1.27	
D	(dithiobenzoate with C(CH$_3$)(CN)CH$_2$CH$_2$CO$_2$H)	MA	$M_n = 35 \times 10^3$ g mol^{-1} PDI = 1.26	[202]
E	(dithiobenzoate with C(CH$_3$)(CN)CH$_2$CH$_2$OH)	MMA	$M_n = 2$ to 17×10^3 g mol^{-1} PDI = 1.26	[192]
F	(dithiobenzoate with CH(CH$_3$)COOH)	N-acryloyl-morpholine	$M_n = 10 \times 10^3$ g mol^{-1} PDI = 1.35	[376]

MA methyl acrylate, *PDI* polydispersity index

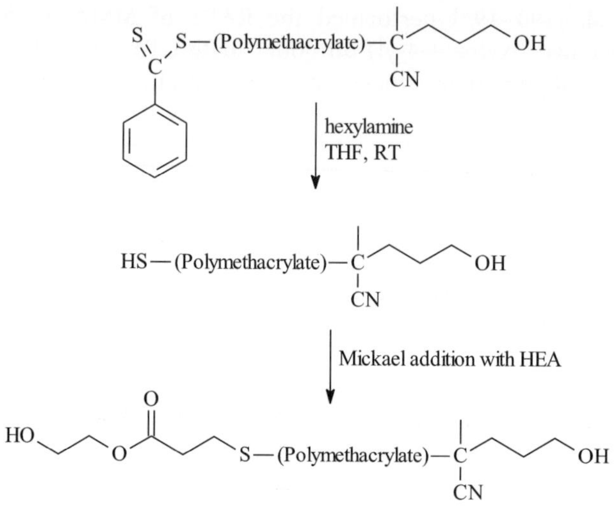

Scheme 33 Synthesis of hydroxy-telechelic poly(methyl methacrylate) (*PMMA*). *HEA* hydroxyl ethyl acrylate, *THF* tetrahydrofuran

To quantify the hydroxy-telechelic PMMA, LC was used to show that this two-step reaction yielded about 67% of telechelic oligomers [192]. This unexpected low yield was explained by several factors: during RAFT some disproportionation may occur leading to dead chains without any terminal thioesters; also some side reactions occurred from aminolysis of the terminal thioester, leading to a hydrogen-terminated chain unable to be functionalized into a hydroxyl group.

Another example is the work of Perrier et al. [202] that proposes first to remove the terminal thioester group (after RAFT) and second to recover the CTA. To achieve the bifunctionality and the recovery of the CTA, the monofunctional oligomer is placed in solution with a high extent of initiator (polymer-to-initiator concentration ratio 1 : 20). The radical provided by the initiator will react on the reactive C = S bond of the terminal thioester. By using an excess of initiator radical, the fragmentation will occur and free the new leaving thioester group, directly replaced by a radical provided by the excess of initiator (Scheme 34).

Scheme 34 illustrates that it is necessary to choose the right initiator, i.e., bearing a further condensable function. This function will correspond to the second end group in the polymer. For instance, Perrier et al. have undertaken the RAFT of methyl acrylate with a dithioester bearing a carboxyl function (entry D in Table 17). The monofunctional oligoacrylate was reacted with an excess of ACVA. It is noteworthy that carboxy-telechelic oligomethyl acrylate was obtained in a one-step reaction. Owing to the ACVA structure, the same transfer agent is recovered at the end of the reaction.

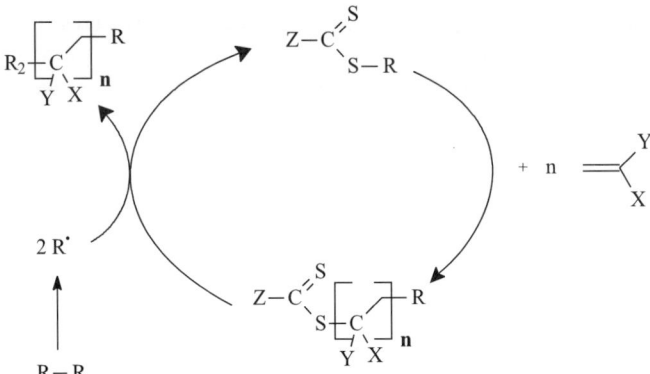

Scheme 34 Reaction cycle to get telechelic polymers and to recover the CTA

2.7
Telechelic Oligomers Obtained by Nitroxide-Mediated Polymerization

The use of nitroxides as mediating radicals has been revealed to be highly successful in living free-radical polymerization and has received considerable attention for more than 10 years [203–207]. Much attention has been devoted to understanding the mechanism (Scheme 35) and kinetics for NMP [208–210].

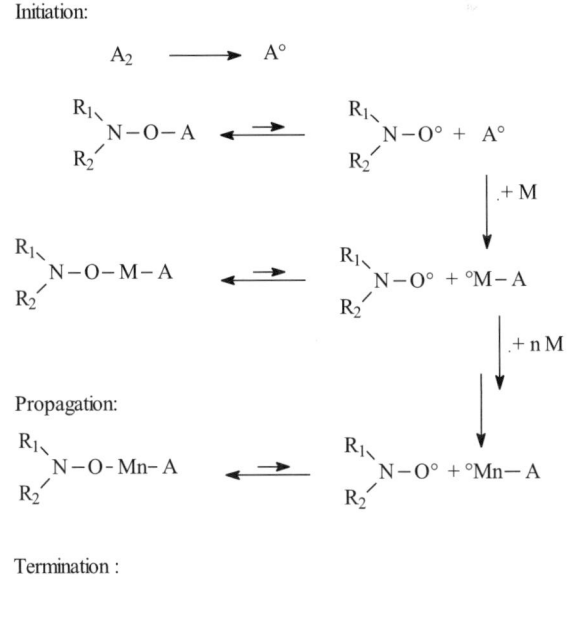

Scheme 35 Nitroxide-mediated polymerization (*NMP*)

NMP was first viable for styrene and its substituted derivatives [211], but was extended to acrylates, acrylamides, and some other vinyl monomers [212]. Hawker et al. [7] reviewed the overall mechanism of NMP as well as several nitroxides and their reactivity.

Recently, research has been focused on synthesizing telechelic oligomers by the use of NMP. α,ω-functionalized polymers may be reached either through a bimolecular process or through unimolecular initiators. The bimolecular process is based on a combination of nitroxide and radical initiator [213, 214]. In that case, the functionality will be gained by using a functional initiator. The ω-extremity of the polymer will, however, remain the aminoxyle function. The bimolecular process was first used in NMP of styrene with benzoyl peroxide as the initiator and TEMPO as the mediator [215]. The alternative procedure for the synthesis of chain-end functionalized polymers relies on the use of active species carrying both the desired functional group and an aminoxyl unit. Like for the bimolecular process, the polymer will carry the aminoxyl function. In both cases, getting the bifunctionality will require a chemical modification of the terminal aminoxyl function (Scheme 36).

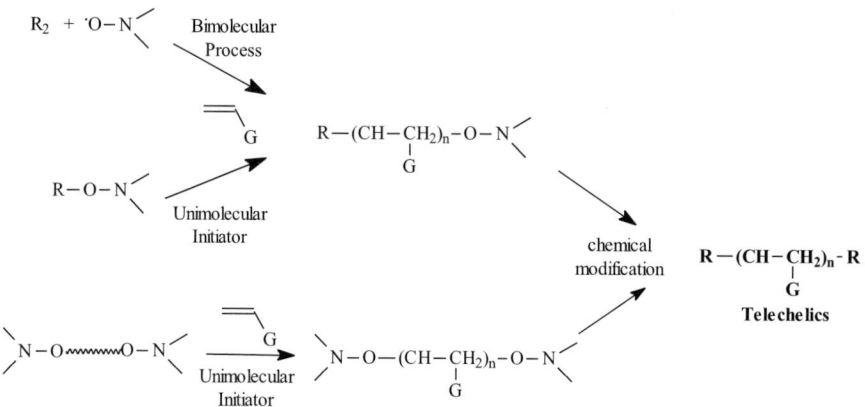

Scheme 36 Synthesis of a telechelic polymer by NMP

2.7.1
Synthesis of Precursors of Telechelic Oligomers

This section is devoted to the synthesis of oligomers, which are not real telechelics, but are able to give telechelics by chemical modification.

Concerning the bimolecular process, we give some examples of their synthesis. The bimolecular process, based on the use of a functional initiator, was not employed much to get telechelic oligomers. The main reason is that, despite the use of a counter radical, it is difficult to avoid any termination reactions occurring by recombination with the initiator radical. It is, however, interesting to outline the work of Pradel et al. [216–218], patented [219] by Elf-Atochem. Pradel et al. [216–218] performed the NMP of 1,3-butadiene initiated by hydrogen peroxide and controlled by the use of TEMPO. Interestingly, they proved, by plotting $\ln([M_0]/[M])$ versus time, that the reaction occurred in two phases. The first phase corresponds to the formation of the monoadduct and then polymerization occurred in a second phase, represented by a straight line [218]. They characterized the expected structure by ^{1}H NMR, especially the methylene protons in the α-position of both aminoxyl and hydroxyl functions.

In a similar way, Hawker and Hedrick [220] synthesized α-amino,ω-aminoxyl PS. Before performing the NMP of styrene, they synthesized a new protected amino diazoic initiator by reaction of N-(*tert*-butoxycarbonyl)-4-aminophenol with a bisacid chloride diazo initiator. The resulting initiator was heated in the presence of styrene and TEMPO at 130 °C (Scheme 37).

The system was proved to be living [204, 213]. The polymer obtained exhibited M_n of 14 000 g mol^{-1} with PDI of 1.2; hence, the authors realized the deprotection of the polymer to end up with a monoamino-terminated PS.

Concerning the unimolecular process, we give firstly some nonexhaustive examples of mononitroxides and binitroxides used in this area (Table 17).

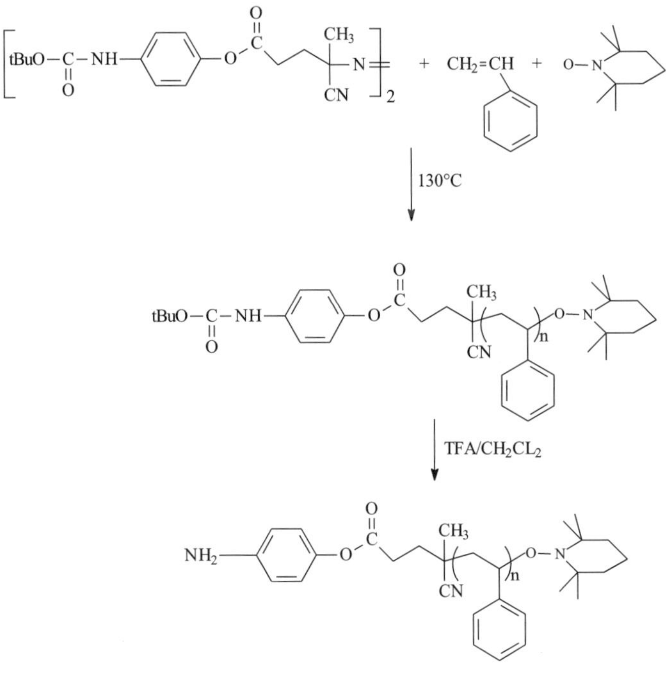

Scheme 37 Synthesis of α-amino,ω-aminoxyle polystyrene by NMP. *TFA* trifluoroacetic acid

Table 17 Some unimolecular initiators used in nitroxide-mediated polymerization to lead to monofunctional oligomers

Entry	Unimolecular initiator	Refs.
A		[220, 344]
B		[220, 344]
C		[378]
D		[378]
E		[206]
F		[377]

Table 17 (continued)

Entry	Unimolecular initiator	Refs.
G		[221]
H		[212, 224, 379]
I		[378]

For instance, Hawker and Hedrick [220] realized the NMP of styrene in the presence of compound B at 130 °C to afford functionalized PS with M_n of 13 500 g mol^{-1} and PDI of 1.16. The *tert*-butyloxycarbonyl protected group, borne by the unimolecular initiator, was replaced by an amino group with trifluoroacetic acid.

Li et al. [221] were interested in getting nitroxide-telechelic PS. They synthesized compound G by double hydrogen atom abstraction from *p*-diethylbenzene in the presence of TEMPO [222]. NMP of styrene was then carried out. ^{13}C NMR confirmed the presence of TEMPO moieties:

These examples are numerous. However the authors, instead of making the chemical modification in order to obtain telechelic oligomers, used these compounds for obtaining diblock or triblock copolymers.

2.7.2
Synthesis of Telechelic Oligomers

Solomon et al. [203] developed a technique allowing the terminal aminoxyl to be replaced by a hydroxyl function. With this aim, they reacted the terminal aminoxyl containing oligomer with acetic acid catalyzed by zinc. Pradel et al. [218] achieved the synthesis of hydroxy-telechelic polybutadiene by applying the methodology of Solomon et al. to α-hydroxyl,ω-aminoxyl polybutadiene (the synthesis was presented earlier) at 80 °C. After 2 h, they obtained a quantitative reduction of the aminoxyl functions evidenced by ^{1}H NMR (Scheme 38). The average hydroxyl functionality of oligobutadiene was 2.06.

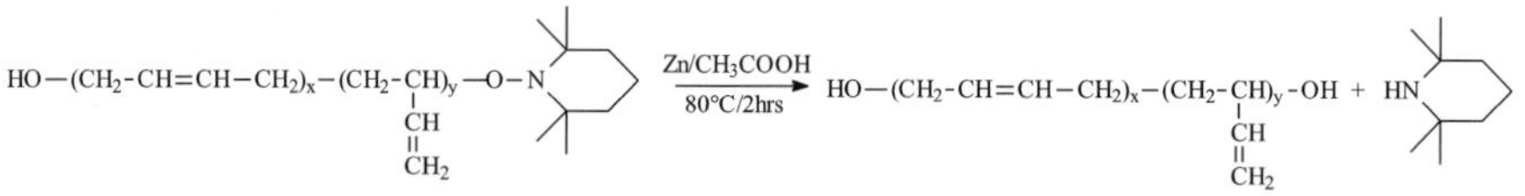

Scheme 38 Synthesis of hydroxy-telechelic polybutadiene

Harth et al. [223] recently developed a new methodology to replace the terminal aminoxyl based on the addition of one single maleic anhydride unit, considering that addition of a second unit is disfavored. To mimic this approach, α-hydrido alkoxyamine (Scheme 39, **1**) was reacted with 2 equiv of N-phenyl maleimide, leading to addition of one unit. Upon heating, the cor-

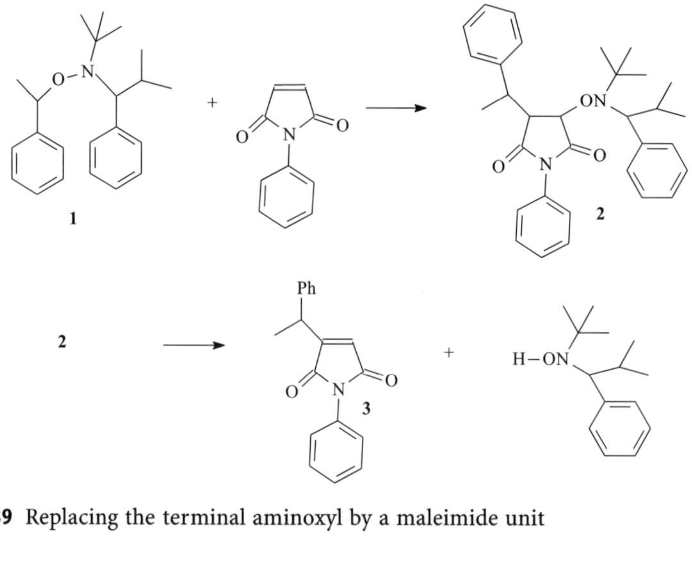

Scheme 39 Replacing the terminal aminoxyl by a maleimide unit

responding product (Scheme 39, **2**) eliminated the terminal-aminoxyl to give the substituted maleimide derivative (Scheme 39, **3**) with more than 90% yield when conducted in DMF.

Using this procedure, Harth et al. [223] synthesized a pyrene-labeled PS by reacting alkoxyamine-terminated PS with 4-pyrenylbutylmaleimide. The molecular weight and PDI of maleimide-terminated PS were similar to those of nitroxide-terminated oligostyrene. The authors however showed that maleimide-terminated PS was more thermally stable than the nitroxide analogue.

We have used an interesting method of coupling oligobutadiene intiated by H_2O_2 and terminated by TEMPO [219]. This method is based on the continuous elimination of the TEMPO unit by sublimation, allowing the reaction equilibrium to be displaced by a simple thermal means:

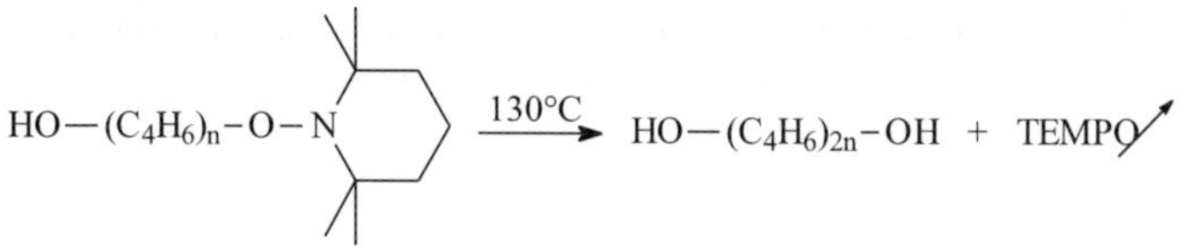

$$HO-(C_4H_6)_n-O-N\diagup \xrightarrow{130°C} HO-(C_4H_6)_{2n}-OH + TEMPO\cdot$$

As can be seen, very few examples of telechelic oligomers have been reported in the literature, although NMP is also a good alternative for obtaining these kinds of compounds. Noteworthy previous studies were realized for the synthesis of telechelic oligomers bearing associative groups at both ends.

With the incorporation methodology of the maleimide unit, Lohmeijer et al. [224] were able to synthesize terpyridine-telechelic PS (Scheme 40, **6**). They utilized a terpyridine-functionalized maleimide (Scheme 40, **5**) that replaced the nitroxide chain end of PS (Scheme 40, **4**). The polymer obtained would be of great value to prepare ABA metallo-supramolecular triblock copolymers.

This "construction" is based on the use of a metal complex, serving as supramolecular linker between blocks. Terpyridine ruthenium was proved to be an efficient linker [225].

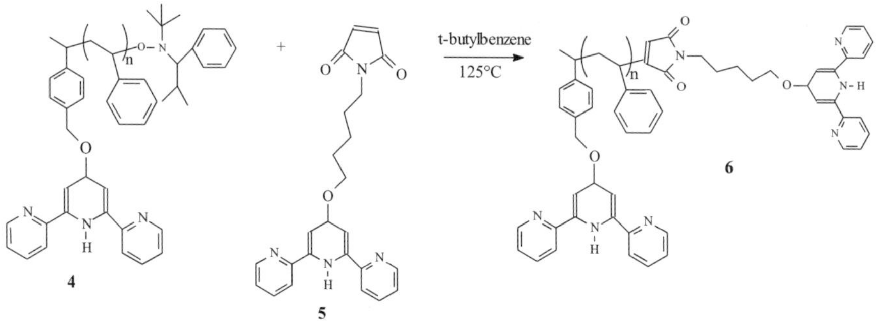

Scheme 40 Synthesis of terpyridine-telechelic polystyrene

2.8
Telechelic Oligomers Obtained by Iodine Transfer Polymerization

Like RAFT, ITP is a degenerative transfer polymerization using alkyl halides [10, 11]. ITP was developed in the late 1970s by Tatemoto et al. [226–229]. In ITP, a transfer agent RI reacts with a propagating radical to form the dormant polymer chain P – I. The new radical R· can then reinitiate the polymerization. In ITP, the concentration of the polymer chains is indeed equal to the sum of the concentrations of the transfer agent and of the initiator consumed. The newly formed polymer chain P·′ can then propagate or react with the dormant polymer chain P – I or R – I [230]. The mechanism of ITP with alkyl iodide is shown in Scheme 41.

More recently, several investigations have shown that ITP can produce telechelic oligomers. The degenerative transfer process then requires the use of diiodide compounds instead of the iodide compounds usually employed in ITP. Noteworthy, the Dupont [231] and Ausimont [232] companies were first attracted by this concept (using IC_4F_8I as the transfer agent) in the CRP of

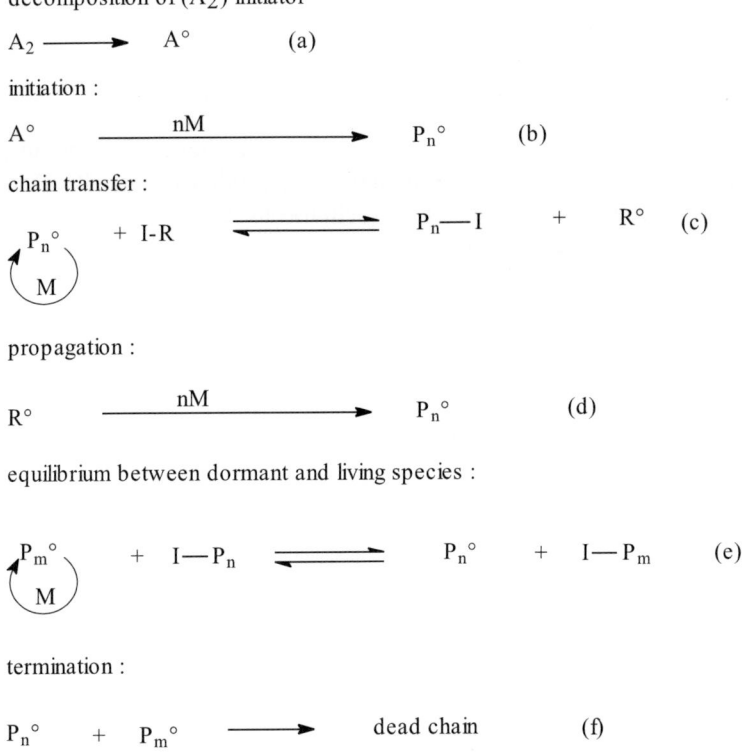

Scheme 41 Elementary steps of iodine transfer polymerization (*ITP*)

fluorinated monomers (i.e., VDF, etc.). In this area, our laboratory has investigated in detail the pseudo living telomerization with these types of iodide compounds. We prepared new oligomers containing various kinds of fluorine units ($CH_2 = CF_2$; C_3F_6 and C_2F_4) to lower the T_g. Taking into account the reactivity and the thermal stability of these oligomers, we find the the best sequence is as follows:

$$I-C_nF_{2n}-I + C_3F_6 \xrightarrow{230°C} IC_3F_6-C_nF_{2n}-C_3F_6I$$
$$\text{A}$$

$$A \xrightarrow{170-180°C} I-(CF_2-CH_2)_x-C_3F_6-C_nF_{2n}-C_3F_6-(CH_2-CF_2)_x-I$$
$$n = 4, 6, 8$$

For $n = 2$ this sequence is not possible because the deiodination of the precursors occurs:

$$IC_2F_4I \longrightarrow I_2 + C_2F_4$$

Percec et al. [233–237] recently reported the synthesis of α,ω-diiodo poly(vinyl chloride) (PVC) by combination of single electron transfer (SET) and degenerative chain transfer (ITP) (Scheme 42).

To get the diiodo telechelic structure, the authors used iodoform and methylene iodide as degenerative CTAs. SET involves the production of radical ions generating free radicals and anions or cations. Iodoform and methylene iodide can be activated both by degenerative transfer mediated by the growing PVC radicals and by SET. The reaction may be catalyzed by sodium dithionite ($Na_2S_2O_4$). These catalyzed reactions allow the suppression of side reactions but also the scavenging of oxygen. The LRP of vinyl chloride resulted in diiodo PVC with M_n of 6000– 10 000 g mol^{-1} and with PDI of about 1.6.

However, the chain-end functionality remained an iodine atom. Further investigations to replace the iodine atom by an interesting polycondensable function were undertaken by various authors.

Several ways are possible to enable the diiodides to be functionalized after ITP has proceeded. Basically, these routes can be gathered into three families (Scheme 43).

We illustrate each route with an example.

2.8.1
Direct Chemical Change

Only a few studies used the direct chemical change of the terminal iodine atom into a further condensable function. To our knowledge, the direct chem-

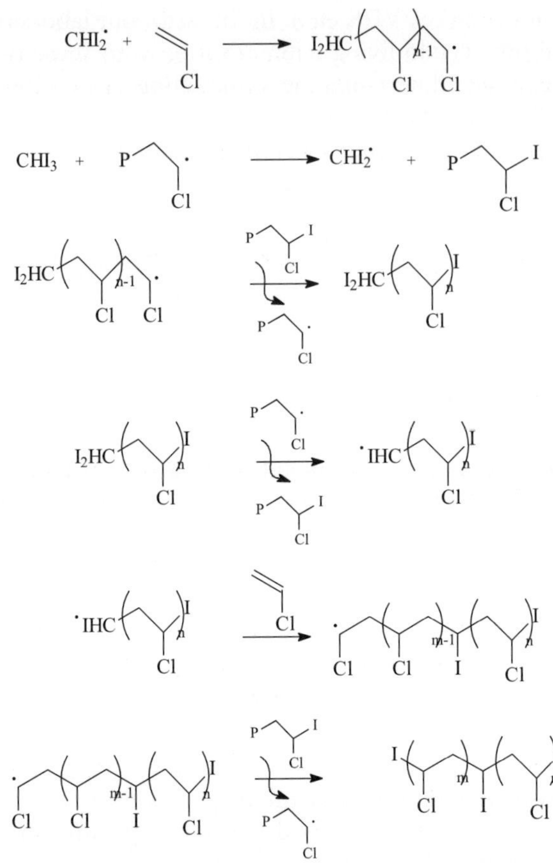

Scheme 42 Synthesis of diiodo poly(vinyl chloride) (*PVC*) by combination of ITP and single electron transfer (*SET*) with iodoform in the presence of $Na_2S_2O_4$

i) Direct chemical change (G being the condensable function):

$$I-(M)_n-I \xrightarrow[R=metal]{2R-G} G-(M)_n-G + 2IR$$

ii) Functionalization from addition of iodide to ω-functional-α-ethylenic derivatives (G and R being the condensable function and the spacer group, respectively):

$$I-(M)_n-I + 2\,H_2C=CH-R-G \longrightarrow G-R-CHICH_2-(M)_n-CH_2CHI-R-G$$

$$G-R-CHICH_2-(M)_n-CH_2CHI-R-G \xrightarrow{H_2 \text{ or } H^+} G-R-C_2H_4-(M)_n-C_2H_4-R-G$$

iii) Radical coupling (G being the condensable function):

$$I-(M)_n-I + 2\,I-G \longrightarrow G-(M)_n-G + 2\,I_2$$

Scheme 43 Chemical modification of chain-end iodide atom

ical change only concerns fluoromonomers, such as VDF, which were polymerized through the ITP process in the presence of diiodide transfer agents. We have summarized these works on direct chemical change in a specific book [238].

Feiring [239] synthesized fluorinated diamine as follows:

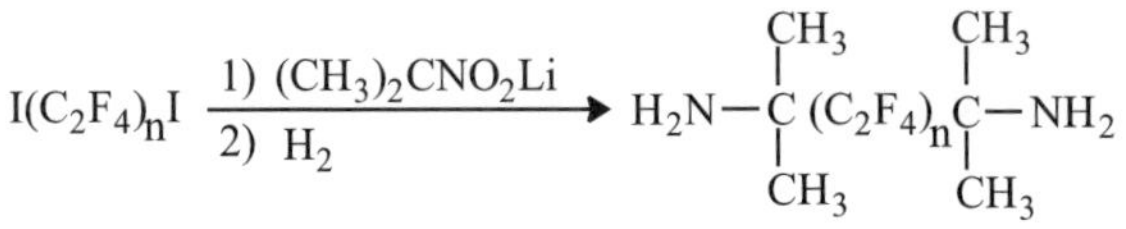

$$I(C_2F_4)_nI \xrightarrow[\text{2) } H_2]{\text{1) } (CH_3)_2CNO_2Li} H_2N-\underset{\underset{CH_3}{|}}{\overset{\overset{CH_3}{|}}{C}}(C_2F_4)_n\underset{\underset{CH_3}{|}}{\overset{\overset{CH_3}{|}}{C}}-NH_2$$

2.8.2
Functionalization by Radical Addition

Interestingly, Percec et al. [234] demonstrated that the chloroiodomethyl chain ends of PVC can be replaced by other functional groups that are further condensable. For instance, PVC was functionalized by SET $Na_2S_2O_4$-catalyzed with 2-allyloxyethanol (Scheme 44). After precipitation, the functionalization resulted in α,ω-hydroxy PVC with 90% yield. The catalytic effect of $Na_2S_2O_4$ first led to the abstraction of the chain-end iodine atom, followed by radical addition of 2-allyloxyethanol. Then the hydrogen abstraction onto 2-allyloxyethanol allowed the hydroxyl-telechelic structure to be obtained.

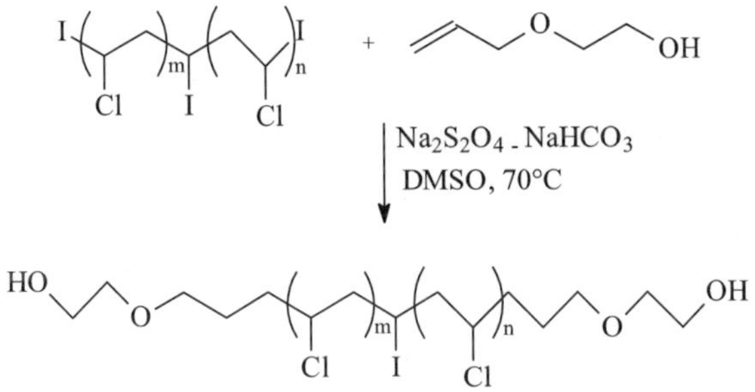

Scheme 44 Synthesis of dihydroxy PVC by SET with 2-allyloxyethanol in the presence of $Na_2S_2O_4$

2.8.3
Radical Coupling

Like the direct chemical change, the radical coupling mainly concerns fluorinated monomers. Amureri and Boutevin [238] summarized the different studies concerning the modification of α,ω-fluoropolymers in a recently published book. They showed, for instance, that extensive research [240] was carried out on the synthesis of diaromatic difunctional compounds linked to fluorinated chains according to the following Ullman coupling reaction:

$$\text{G}-\!\!\left\langle\bigcirc\right\rangle\!\!-\text{I} + \text{I}-\text{R}_\text{F}-\text{I} \xrightarrow[\text{DMSO}]{\text{Cu}} \text{G}-\!\!\left\langle\bigcirc\right\rangle\!\!-\text{R}_\text{F}-\!\!\left\langle\bigcirc\right\rangle\!\!-\text{G}$$

where G represents a functional group, such as hydroxyl (e.g., bisphenol), carboxylate, isocyanate [241], or nitro (precursor of amine) in *para* and *meta* positions towards the fluorinated chain. McLoughlin and Thrower [242, 243] also attached several functional groups onto each aromatic ring, in 85% yield, such as the following tetracarboxylates:

$$\text{EtO}_2\text{C}\!\!-\!\!\left\langle\bigcirc\right\rangle\!\!-\!(\text{CF}_2)_n\!\!-\!\!\left\langle\bigcirc\right\rangle\!\!-\text{CO}_2\text{Et}$$
(with additional EtO_2C and CO_2Et substituents)

From these compounds, Critchley et al. [158] prepared novel polymers such as polyesters, silicones, and polyimides.

In a similar way, our team has done lots of work in functionalizing α,ω-diiodoperfluoroalkanes into fluorotelechelic compounds. These works were summarized by Ameduri et al. [244] in a review on the synthesis of fluoropolymers. For instance, our team synthesized α,ω diols or dienes of perfluoroalkanes [245–248]. These compounds are precursors of hybrid fluorosilicones [249] but also of thermoplastic elastomers by polycondensation with polyimide sequences [250].

3
Synthesis of Macromonomers by Radical Techniques

As already mentioned in the "Introduction," macromonomers can be considered as precursors of graft copolymers, whereas telechelic oligomers will lead to multiblock copolymers.

Graft copolymers are generally obtained by using one of the following three methods:

1. "Grafting onto", corresponding to the attachment of side chains to the backbone
2. "Grafting from", corresponding to the side chains grafted from the backbone
3. "Grafting through", involving the copolymerization of macromonomers (made either from other living methods or from conventional radical methods using other small monomers)

This second part is devoted to the last method. Several books and reviews, such as Rempp and Franta [3], focused on macromonomers; however these publications are not recent and a new review of the current status of macromonomers is necessary. Indeed these last few years have witnessed the development of techniques such as addition–fragmentation and CCT, and also the use of LRP, leading to new macromonomers. Moreover, anionic polymerization, polycondensation, ring-opening polymerization, and coordination polymerization have given original structures that can be (co)polymerized by a radical route. In the first section an overview of such macromolecules is given. But, in order to give a general view on the area of macromonomers, we will mainly describe the synthesis of these compounds by all the radical techniques. Before concluding, the reactivity of macromonomers will be enhanced.

3.1
New Macromolecular Designs of Macromonomers

Most macromonomers are made from the macromolecular chain linked to a reactive double bond for further radical polymerizations.

3.1.1
Acrylic and Styrenic Double Bonds

The reactive double bonds are usually either acrylic or styrenic types. These macromonomers can be classified in three categories (Scheme 45).

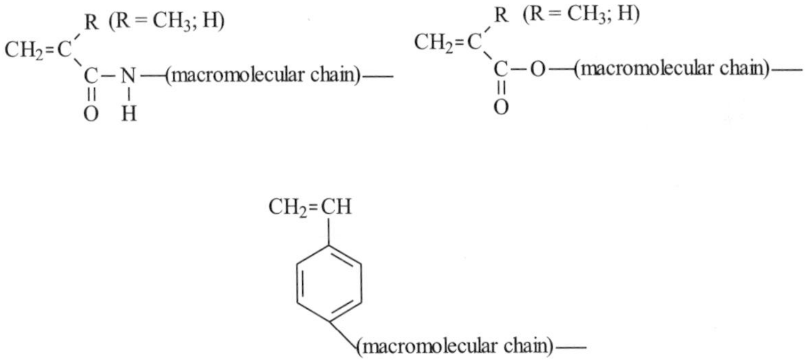

$$CH_2=C\begin{smallmatrix}R\ (R=CH_3;\ H)\\ \\ C-N-(macromolecular\ chain)-\\ \|\ \ |\\ O\ \ H\end{smallmatrix} \qquad CH_2=C\begin{smallmatrix}R\ (R=CH_3;\ H)\\ \\ C-O-(macromolecular\ chain)-\\ \|\\ O\end{smallmatrix}$$

$$CH_2=CH-C_6H_4-(macromolecular\ chain)-$$

Scheme 45 Structures of macromonomers bearing a vinyl group

The originality and the specificity of the macromonomer structure is provided by the macromolecular chain. In this section we are going to illustrate such specificities by some relevant examples for each type of macromonomer described before. The method for obtaining such macromonomers is also given. The A-type macromonomer is usually an amido-type for the vinyl group (Table 18). Tables 18 and 19 also give some examples of acrylic- and styrenic-type macromonomers, respectively. Obviously, the C-type macromonomers, which bear a polymerizable styrenic group, are the most synthesized ones (Table 20).

Table 18 Some examples of A-type macromonomers and their applications

Macromolecular chain	Synthesis of macromonomers	Monomer(s) copolymerized	Applications	Refs.
I—(CH$_2$)$_3$—NH—C—CF$_2$—(O—CF$_2$—CF)$_{7\text{-}5}$—CF$_3$ ‖ | O CF$_3$	/	/	Biofouling	[380, 381]

O
    ‖
I—(N—C)$_n$—
    |
    C$_6$H$_{13}$
	TiCl$_3$OCH$_2$CF$_3$ coordination	MMA	/	[382, 383]

I—(CH$_2$)$_3$-O—(CH$_2$-CH$_2$O)$_n$—(CH$_2$)$_3$=NH—(CH$_2$)$_3$-SO$_3$H	Chemical modification of diaminoPEG	Dimethyl-PEG	Hydrogels	[384]

R
              |
I—(CH$_2$)$_n$-NH—(C—CH—NH)$_n$-H
              ‖
              O
		Acrylamide	Thermotropic and lyotropic	[385]

PEG poly(ethylene glycol)

I is the unsaturated group CH$_2$=C$\overset{R}{\underset{\underset{O}{\overset{\|}{C}}-\underset{H}{\overset{\|}{N}}-}{}}$

Table 19 Some examples of B-type macromonomers and their applications

Macromolecular chain	Synthesis of macromonomers	Monomer(s) copolymerized	Applications	Refs.
$I-C_2H_4-(O-\overset{CH_3}{\underset{C=O}{C}}-CH_2-CH_2)_n-O-\overset{CH_3}{\underset{\|}{C}}-CH=CH$	1. PHB depolymerization 2. Esterification with HEA	MMA	–	[386, 387]
$I-C_2H_4-(O-\overset{CH_3}{\underset{C=O}{C}}-CH_2)_n-O-\overset{CH_3}{\underset{C=O}{C}}-CH-OH$	MHEA + Al(OR)$_3$ + lactide	MMA, NVP NIPAM, MA	Biocompatibility	[388–391]
$I-(C_4H_6-O)_m-H$	Methacryloic chloride + AgClO$_4^-$	Sty	–	[392, 393]
$I-(CH_2-CH_2O)_n$-(substituted phenyl group)	–	AA	–	[394]
$I-C_2H_4-NH-\overset{O}{\underset{\|}{C}}-O-C_2H_4-S-(CH_2-\overset{CH_3}{\underset{CO_2CH_3}{C}}{}_h-H$	Telomerization	Glycidyl acrylate	–	[254]
-Lignin	–	MMA	Biodegradability	[395, 396]

I is the unsaturated group: $CH_2=C\overset{R}{\underset{C=O}{\diagdown}}-O-$

Table 20 Some examples of C-type macromonomers

Macromolecular chain	Synthesis of macromonomers	Monomer(s) copolymerized	Applications	Refs.
(glucosamine-maltopentaose structure with I–CH₂–NH–C(=O)–)	+ maltopentaose-CH₂NH₂ (4-vinylbenzylamine)	Sty	Biomedical material	[397, 398]
I–CH₂–(CH₂–CH)ₙ–(CH₂–C(CH₃))ₘ– (polystyrene chain)	Anionic polymerization	–	Cylinder brushes	[399, 400]
(styrenesulfonate with TEMPO-alkoxyamine, SO₃Na)	Nitroxide functionalization	Sty	Proton exchange membranes	[342, 401]
I–CH₂–O–CH₂CH₂S–(CH₂–CH₂)ₙ–H, C=O, NH, CH(CH₃)₂	Telomerization	Sty	Thermosensitivity	[257]

I is the unsaturated group: CH₂CH–C₆H₄–

3.1.2
Other Reactive Double Bonds

Although acrylic and styrenic bonds are the most common double bonds of the macromonomers, some authors were interested in introducing unusual vinyl groups. Table 21 gathers some macromonomer structures bearing such peculiar reactive double bonds.

Table 21 Peculiar reactive double bonds of macromonomers

Macromonomer structure	Refs.
$CH_2=CH-O-C(=O)-CH_2-(CH_2-CH(C_6H_5))_n-Cl$	[318]
$CH_2=CH-O-C_2H_4-O-C(=O)-C(CH_3)_2-(CH_2-C(CH_3)(CO_2R))_n-Br$	[402]
$CH_2=CH-CH_2-NH-C(=O)-CH_2-(CH_2-C(CH_3)(CO_2C_2H_4-N(CH_3)_2))_n-Br$	[319]
$CH_2=CH-CH_2-CH(C_6H_5)-CH_2-(CH(C_6H_5)-CH_2)_n-CH(C_6H_5)-CH_3$	[325]
$CH_3O-(CH_2-CH_2O)_n-C(=O)-CH=CH-C(=O)-(OCH_2-CH_2)_n-OCH_3$	[403]
norbornene-C(=O)-O-C(CH_3)(C_6H_5)-(CH(SBu)-CH_2)_n	[404, 405]

3.1.3
Macromonomers with Polycondensable Groups

As already mentioned, the vinyl group is not the only kind of reactive function found for the macromonomers. A macromonomer can be constituted of two condensable functions at the same chain end. The synthesis of this kind of macromonomer is quite recent, explaining the low number of publications. However, this new category of macromonomers is of great interest because the condensable functions are numerous (hydroxyl, carboxyl, amine, etc.). Table 22 gives some macromonomer structures, i.e., two condensable functions bearing a macromolecular chain.

This section shows that macromonomers exhibit different structural designs. Their reactive group can be either a polymerizable double bond or two polycondensable groups such as hydroxyl groups. The structures of the macromonomers are actually numerous owing to the type of the macromolecular chain. Furthermore, the macromolecular chain of the macromonomer will bring the specific properties of the graft copolymer, obtained after copolymerization of the macromonomer with a conventional monomer. For instance, a lignin-terminated MMA macromonomer will afford biodegradability to a graft copolymer obtained by radical copolymerization of the macromonomer with MMA.

Unlike vinyl-type macromonomers, studies concerning macromonomers bearing two polycondensable functions are rather rare. This may be due to the rather complex syntheses of such macromonomers. In the following sections, we will describe the different methods for synthesizing both vinyl-type and polycondensable-type macromonomers.

3.2
Macromonomers Obtained by Telomerization

Rempp and Franta [3] described the synthesis of macromonomers either by using redox catalysis with halogenated monomers (vinyl chloride, vinyl dichloride, or even trifluorochloroethylene) or by using a radical initiation with (meth)acrylates. In the latter case, thiol compounds were used as transfer agents (Scheme 46) [251]:

$$CH_2=C\overset{CH_3}{\underset{\underset{O}{\overset{\|}{CO}}-CH_2-CH(OH)-CH_2-O-\underset{O}{\overset{\|}{C}}-CH_2-S-(M)_n-H}{}}$$

M = MMA; AMA

We will focus on recent developments made on the synthesis of macromonomers through the telomerization process. This may concern the synthe-

Table 22 Some examples of macromonomers bearing two condensable groups

Macromonomer structures	Synthesis of macromonomers	Monomer(s) copolymerized	Applications	Refs.
HO–CH$_2$–CH–OH 　　　　｜ 　　　　CH$_2$ 　　　　｜ 　　　　S–(CH$_2$–CH)$_m$–H 　　　　　　　｜ 　　　　　　　C=O 　　　　　　　｜ 　　　　　　　OBu	Telomerization	–	–	[279, 406]
(structure with CH$_3$, O–(CH$_2$–C$_h$)$_n$–H, CO$_2$CH$_3$, and diol chain with two OH groups)	Boran complex	–	–	[336]
OH ｜ CH$_2$–CH–CH$_2$–S–(CH$_2$–CH$_m$)–(CH$_2$–CH)$_p$–H ｜　　　　　　　　　　　｜ OH　　　CO$_2$C$_2$H$_4$Rf　CO$_2$–(CH$_2$)$_3$–Si(OSi(CH$_3$)$_3$)$_3$	Telomerization	Toluene-2,4-diisocyanate	Surface properties	[290]
Br-substituted aryl with CH$_2$–(CH$_2$–CH$_n$)–Br side chain	ATRP (Pd complex)	–	Light-emitting diodes	[345]
Aryl–O–CH$_2$–(CH$_2$–CH$_n$)–Br with two ClOC groups	ATRP	(OH)$_2$-PEO	–	[340]
Oxazoline–~~~(CH$_2$–CH$_m$)–Br	ATRP	–	–	[335]

ATRP atom transfer radical polymerization

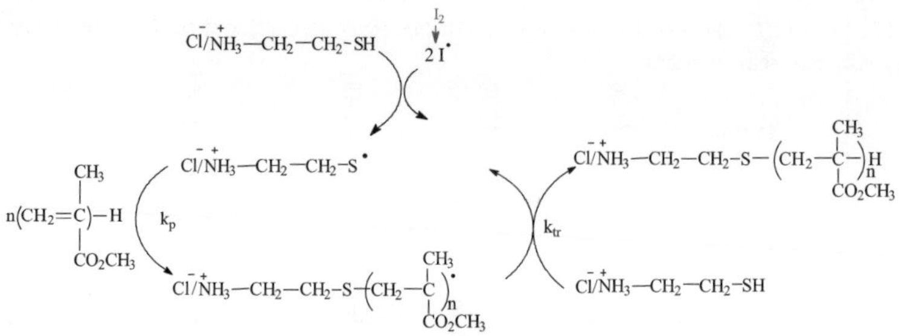

Scheme 46 Mechanism of the telomerization process (example with 2-aminoethanethiol chlorhydrate)

sis of macromonomers bearing either a polymerizable double bond situated in ω-position or two condensable groups situated at the same chain-end.

3.2.1
Macromonomers with a Polymerizable Double Bond

3.2.1.1
Based on New Transfer Agents

Teodorescu [252] developed a direct method for obtaining a polymerizable double bond. The reaction is described in Scheme 47.

The characteristics, i.e., functionality, M_n, and conversion, of VAc macromonomers are given in Table 23.

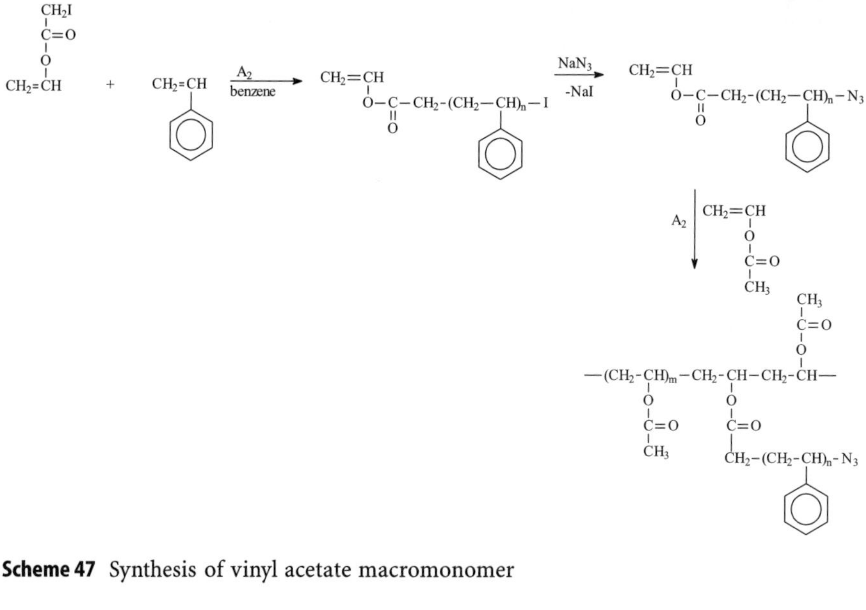

Scheme 47 Synthesis of vinyl acetate macromonomer

Table 23 Characteristics of the macromonomers prepared by iodine transfer polymerization at various reaction times

Runs	Time (h)	Conversion (%)	$M_{n,SEC}$ (g mol^{-1})	PDI SEC	f
VP-St	10	15	4850	1.61	0.93
VP-S	20	27	4610	1.62	0.92
VP-St	30	37	4520	1.65	0.90
VP-St	40	44	4500	1.65	0.88

SEC size-exclusion chromatography

To allow the further copolymerization of this macromonomer with other monomers, the chain-end iodine is extracted by nucleophilic substitution with sodium azide (NaN$_3$).

Our team also realized the synthesis of macromonomers with a polymerizable double bond by using peculiar transfer agents. For instance, telomerizations of (meth)acrylates were performed in the presence of cysteamine, i.e., thiol with an amine group, leading to PMMAs with an amine at the chain end [253]. However, amines enable the Michael addition onto the double bond activated by the carbonyl group. Hence, before performing the telomerization, the amine group is protected (chlorhydrate salt) and recovered by a simple basification of the solution (Scheme 48).

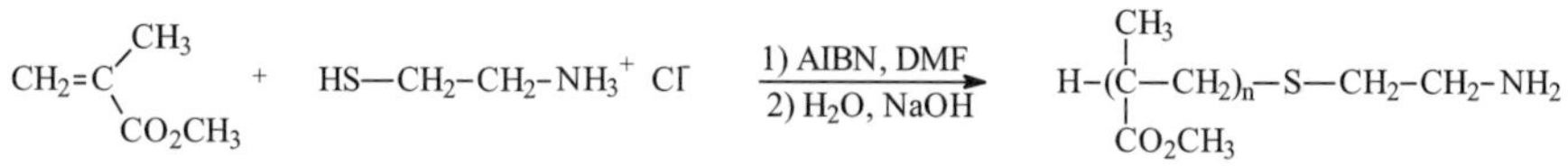

Scheme 48 Synthesis of α-amine poly(methyl metacrylate)

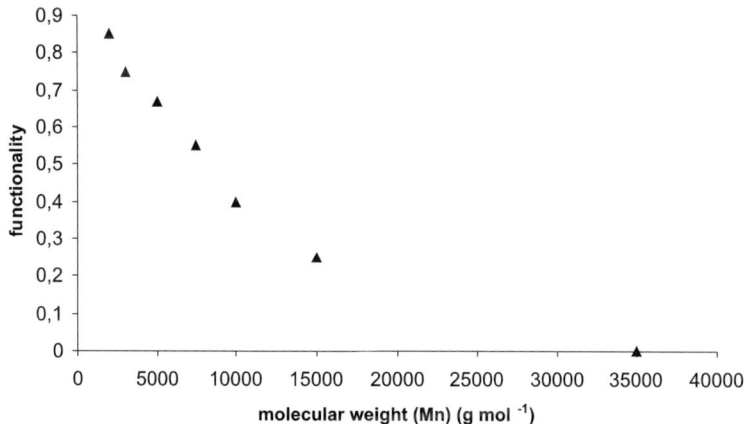

Fig. 3 Amine functionality vs. M_n of poly(methyl methacrylate)

The amine functionality of the polymers was studied versus the molar mass and it is shown to decrease when the molar mass increases (Fig. 3).

The polymerizable double bond may be obtained by functionalizing the amine with a monomer such as methacrylate glycidyl ether or isocyanoethyl methacrylate (IEM), i.e., reaction of amine with an epoxy or isocyanate group. We chose to functionalize the amine by using maleic anhydride to get highly stable (thermally) maleimide macromonomers [162], as shown in Scheme 49.

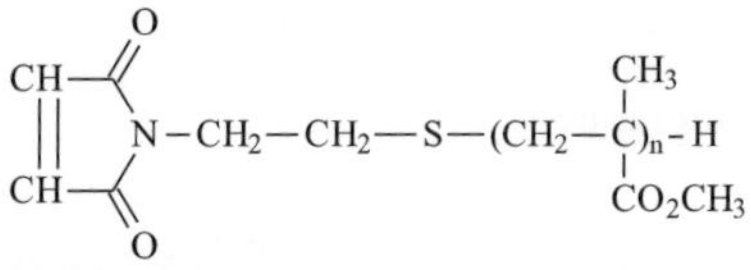

$$\text{CH}=\!\!\!\!\bigg\langle\!\!\begin{array}{c}\text{O}\\\\\text{O}\end{array}\!\!\text{N}-\text{CH}_2-\text{CH}_2-\text{S}-(\text{CH}_2-\overset{\overset{\displaystyle \text{CH}_3}{|}}{\underset{\underset{\displaystyle \text{CO}_2\text{CH}_3}{|}}{\text{C}}})_n-\text{H}$$

Scheme 49 Structure of maleimides macromonomers

3.2.1.2
Based on New Unsaturations

Owing to their isocyanate group, IEM and 1-(isopropenylphenyl)-1,1-dimethylmethyl isocyanate (TMI) have been extensively used to achieve a macromonomer structure.

- *IEM*. Chen et al. [254] developed the process shown in Scheme 50.
  We realized a similar result with dimethylaminoethyl methacrylate as the monomer. Telomerization was performed with 2-mercaptoethanol in the presence of AIBN. This study represents the first example of telomerization of a monomer with a tertiary amine. We showed that the telomerization of such a monomer was highly dependent on the solvent. Indeed, polar solvents strongly favor the nucleophilic addition of thiol onto the methacrylate double bond [255] (Scheme 51). In a second step, oligomers bearing an alcohol group at the chain end can react with IEM to lead to a macromonomer with a methacrylic function.

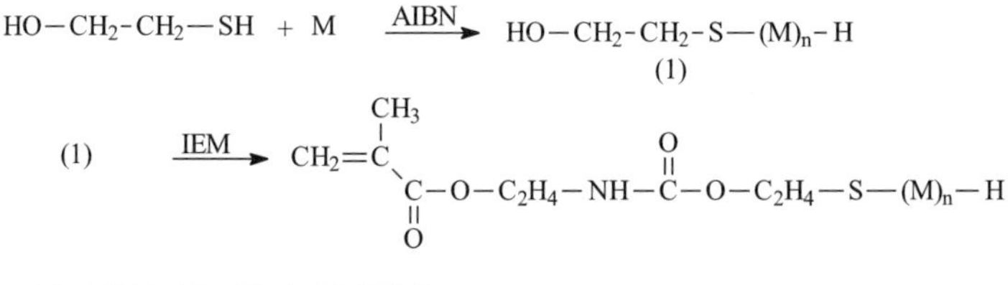

$$\text{HO}-\text{CH}_2-\text{CH}_2-\text{SH} + \text{M} \xrightarrow{\text{AIBN}} \text{HO}-\text{CH}_2-\text{CH}_2-\text{S}-(\text{M})_n-\text{H}$$
(1)

$$(1) \xrightarrow{\text{IEM}} \text{CH}_2=\!\!\overset{\overset{\displaystyle \text{CH}_3}{|}}{\underset{\underset{\displaystyle \text{O}}{||}}{\text{C}}}\!\!-\text{C}-\text{O}-\text{C}_2\text{H}_4-\text{NH}-\overset{\overset{\displaystyle \text{O}}{||}}{\text{C}}-\text{O}-\text{C}_2\text{H}_4-\text{S}-(\text{M})_n-\text{H}$$

M = MMA, ABu, Magly, DMAEMA.

Scheme 50 Synthesis of macromonomers from isocyanoethyl methacrylate. *MMA* methyl methacrylate, *Magly* methacrylate glycidyl ether, *DMAEMA* 2-(dimethylamino)ethyl methacrylate

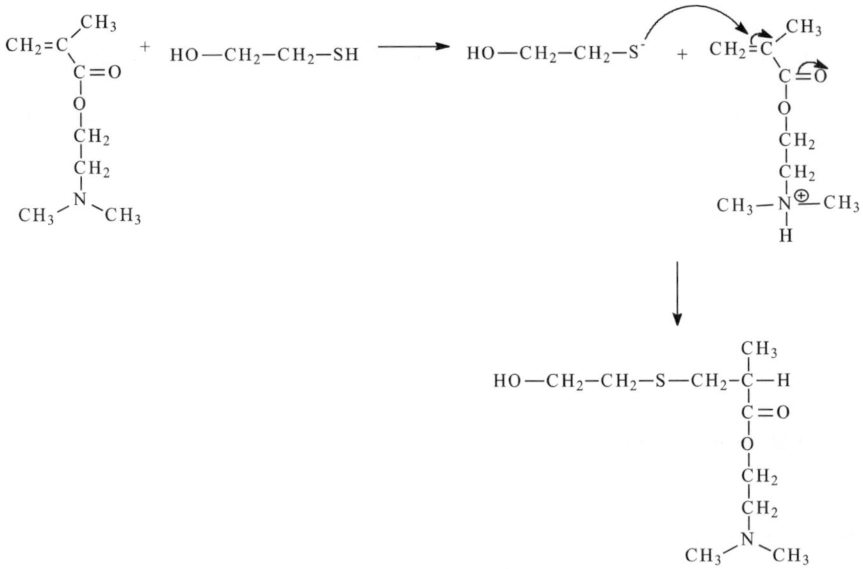

Scheme 51 Nucleophilic addition of thiol onto the methacrylate double bond

In the same way, Oishi et al. [256] used IEM to functionalize oligomers carrying an acid function obtained by polymerization of chiral acrylamides. Chiral polyacrylamide macromonomers were synthesized from 2-methacryloyloxyethyl isocyanate and prepolymers, i.e., poly[(S)-methylbenzyl acrylamide] or poly(L-phenylalanine ethylester acrylamide) with a terminal carboxylic acid or hydroxyl group. Radical homopolymerizations of polyacrylamide macromonomers were carried out under different conditions to obtain the corresponding optically active polymers, as shown in Scheme 52.

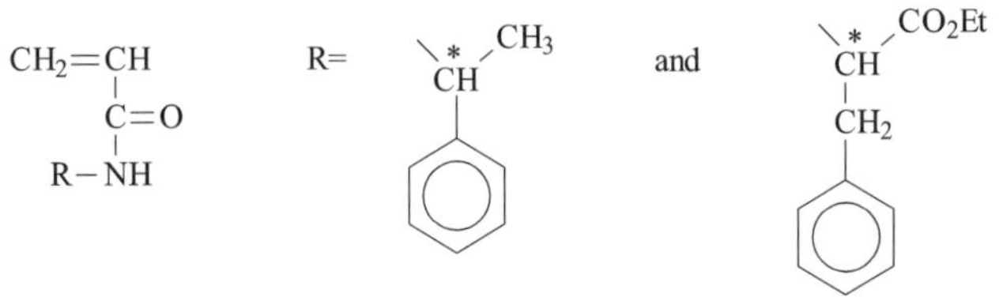

Scheme 52 Structures of (S)-methylbenzyl acrylamide and L-phenylalanine ethylester acrylamide

- *TMI.* Boyer et al. [255] directly used this monomer for the synthesis of N,N'-dimethylethylamino methacrylate macromonomer (Scheme 53).

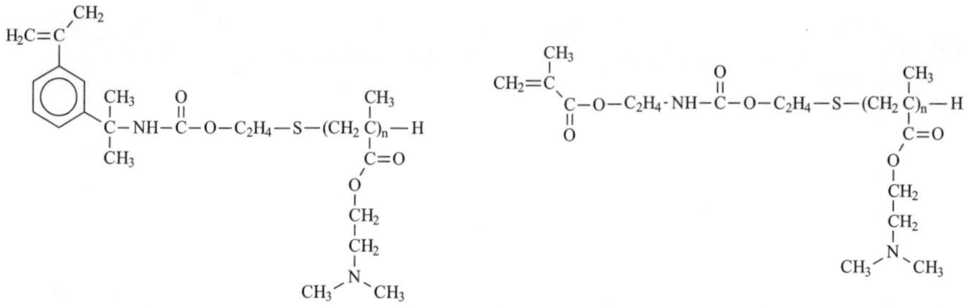

Scheme 53 Macromonomers of N,N'-dimethylethylamino methacrylate obtained by telomerization

3.2.1.3
Based on New Monomers

The group of Akashi [257–263] extensively used NIPAM to synthesize macromonomers with a benzyl group. First, the telomerization of NIPAM is realized with 2-mercaptoethanol, followed by etherification of the alcohol group by using chlorostyrene (Scheme 54).

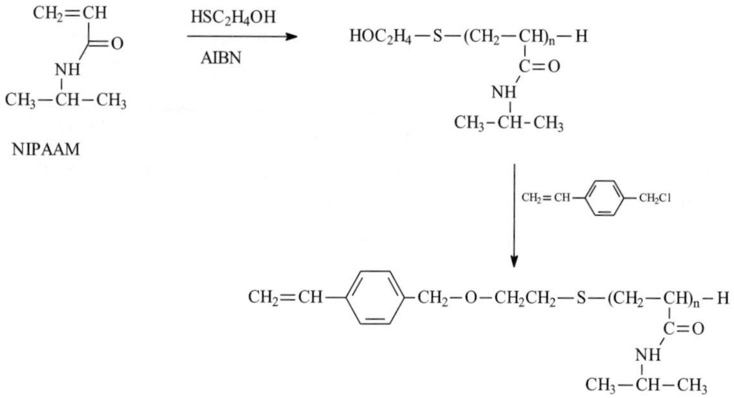

Scheme 54 Synthesis of N-isopropylacrylamide macromonomer

The macromonomers were then copolymerized with styrene in ethanol. The resulting microspheres, with a PS core and poly(NIPAM) brushes, were thermosensitive [264–267].

Another original macromonomer, based on VAc, afforded interesting properties; however the functionalization of such a monomer remains difficult [252, 268–270]. The group of Sato [269, 271–275] suggested the synthesis of VAc macromonomer by functionalization onto VAc telomers, obtained with 2-mercaptoethanol. The functionalization can be realized with acryloyl chloride, giving macromonomers with different molar masses (Table 24).

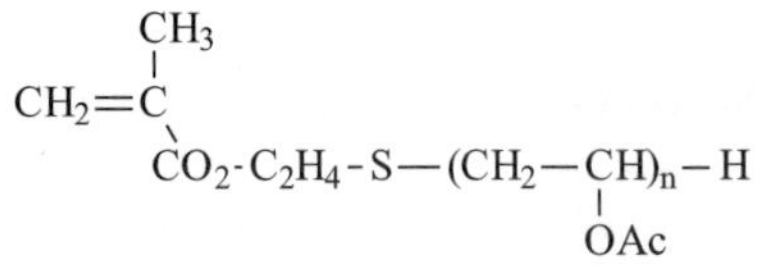

$$CH_2=C(CH_3)-CO_2-C_2H_4-S-(CH_2-CH(OAc))_n-H$$

Table 24 Characterization of poly(vinyl acetate) macromonomers [269]

Samples	DP_n (NMR)	DP_n (SEC)	PDI	Functionality[a]	Functionality[b]
1	20	16	1.9	0.98	0.89
2	31	25	2.1	0.99	0.99
3	57	49	2.9	0.96	0.86
4	107	117	3.1	0.89	0.62

[a] Functionality of hydroxyl group
[b] Functionality of a reactive double bond

In a similar way, Wood and Cooper [268] used isopropoxyethanol as a transfer agent and the telomers were then functionalized with methacryloyl chloride; however, only 28% of the telomers were functionalized with methacryloyl chloride. Macromonomers were then copolymerized with styrene in dispersion copolymerization, in the presence of 1,1,2,2-tetrafluoroethylene. Such copolymers have been used as dispersing agents for the styrene polymerization in supercritical CO_2.

To increase the amount of functionalization, Collins and Rimmer [276] and Carter et al. [277] used 2-propanol as the transfer agent [276, 277]. Owing to the low transfer constant of 2-propanol, a large excess of transfer agent was used. Despite this large excess, they obtained high molar masses (M_n = 10 000 to 17 000 g mol^{-1}).

To increase the alcohol functionalization, telomers of VAc were synthesized with 2-mercaptoethanol in the presence of 2-propanol as the solvent and also with an initiator bearing alcohol groups [276–278] (Scheme 55). Such oligomers were copolymerized with polylactone, leading to poly(vinyl acetate)-polylactone block copolymers.

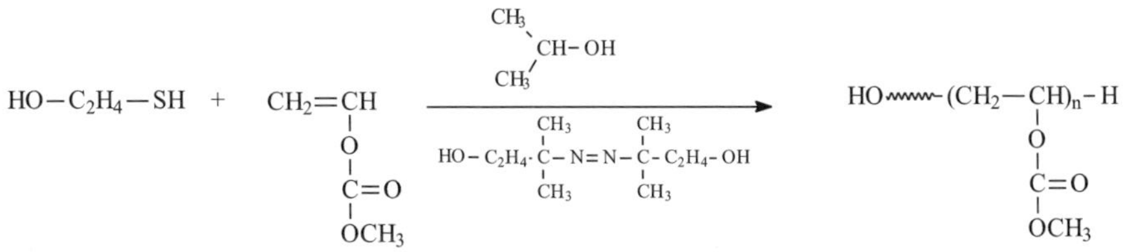

Scheme 55 Poly(vinyl acetate) oligomers bearing an alcohol chain end

3.2.2
Macromonomers with Polycondensable Groups

Several authors performed telomerizations in the presence of transfer agents bearing polycondensable groups. For instance, Nair et al. [279] realized the telomerization of butylacrylate with 1-mercapto-2,3-propanediol, according to Scheme 56.

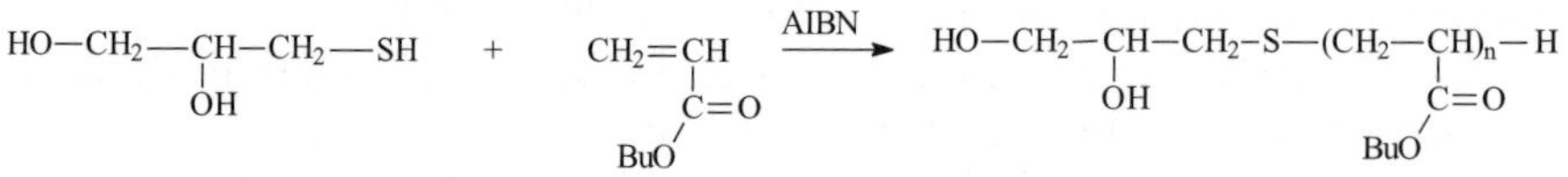

Scheme 56 Synthesis of poly(n-butyl acrylate) macromonomer

1-Mercapto-2,3-propanediol exhibits a transfer constant of 0.55 [279], close to that of 2-mercaptoethanol. The resulting macromonomers, having molecular masses ranging from 2000 to 5000 g mol^{-1}), have been used in polyurethane formulations with poly(oxyethylenes) [280–282].

Similarly, dicarboxylic transfer agents were used in telomerization. For instance, Yamashita [283–288] realized the synthesis of MMA macromonomer by using thiomalic acid as a transfer agent (Scheme 57).

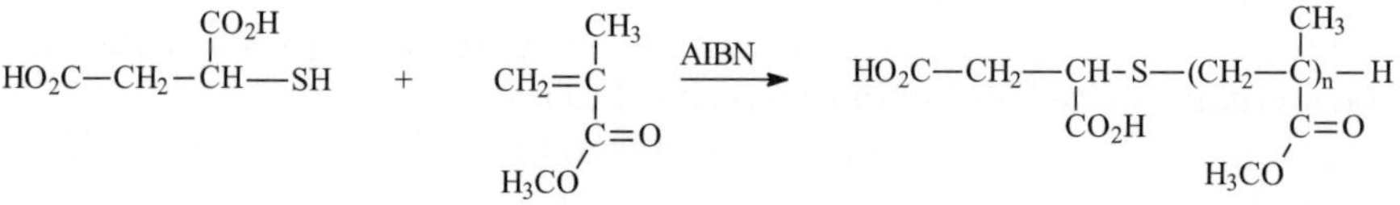

Scheme 57 PMMA macromonomer with two acid groups

Okamoto [289] synthesized identical macromonomers that had been polycondensed with prepolycarbonates, obtaining polycarbonate-*graft*-PMMA. The PMMA grafting chain brings transparency and toughness to polycarbonate matrices. Other authors used this technique to synthesize dihydroxy PS macromonomers, used in the synthesis of polyester by polycondensation with terephtalic acid and butylene glycol.

More complex macromonomers, based on dihydroxy groups, were synthesized [290, 291]:

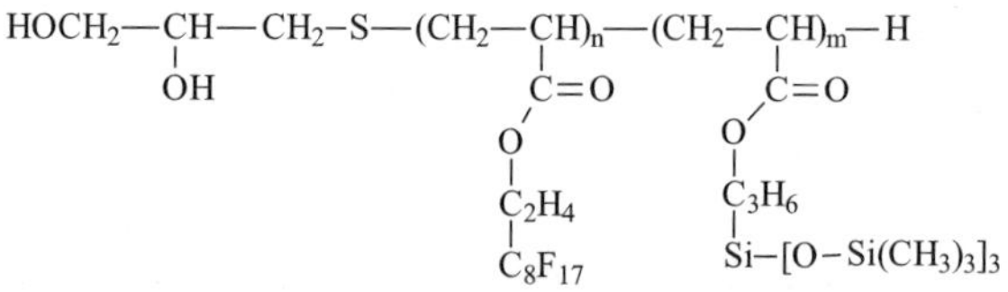

Such a macromonomer was utilized in the synthesis of polyurethanes as adhesives, with very low surface tension (9–12 dynes cm^{-1}), mainly for PVC by using very low rates (1% w/w) [290].

The telomerization process is certainly an appropriate technique for the synthesis of (meth)acrylic-type macromonomers, but also of (meth)acrylonitrile-derived macromonomers. This technique is open to almost all conventional monomers but is also devoted to original monomers such as NIPAM. We can note that macromonomers based on halogenated monomers were synthesized by telomerization, using redox catalysis.

The telomerization technique is essentially based on the use of thiols as transfer agents, bearing a reactive group (hydroxyl, acid, amino). In most cases this reactive group allows the further reactive double bond of the macromonomer to be obtained. Recently, some other transfer agents, based on iodinated compounds, were used to achieve a macromonomer structure. Finally, by using transfer agents bearing two polycondensable groups, telomerization allows the synthesis of macromonomers in only one-step synthesis.

3.3
Macromonomers Obtained by Addition–Fragmentation and Catalytic Chain Transfer

The synthesis of numerous macromonomers can be performed by two methods through radical polymerization in the presence of various addition–fragmentation CTAs [292–295] or catalytic CTAs [69, 70, 296].

3.3.1
Addition–Fragmentation Process

The general form [56] for CTAs involved in addition–fragmentation for the synthesis of macromonomers is described in Scheme 58. The CTA will undergo a β-scission to lead to the corresponding macromonomer (Scheme 59).

Table 25 shows some CTAs involved in addition–fragmentation, leading to the expected macromonomer structure.

Various macromonomers made from an addition–fragmentation process have been employed as precursors of graft copolymers [292, 297–300]. But Krstina et al. [301] also characterized the use of such macromonomers in the synthesis of block copolymers. They explained that for macromonomers based on methacrylic monomers (Scheme 60, 1), fragmentation of the adduct (Scheme 60, 2) (formed by addition of the methacrylate monomer onto the methacrylate macromonomer) always dominates over reaction with the monomer. This fragmentation leads to block copolymers and graft copolymerization does not occur.

The proposed mechanism of block formation is given in Scheme 60.

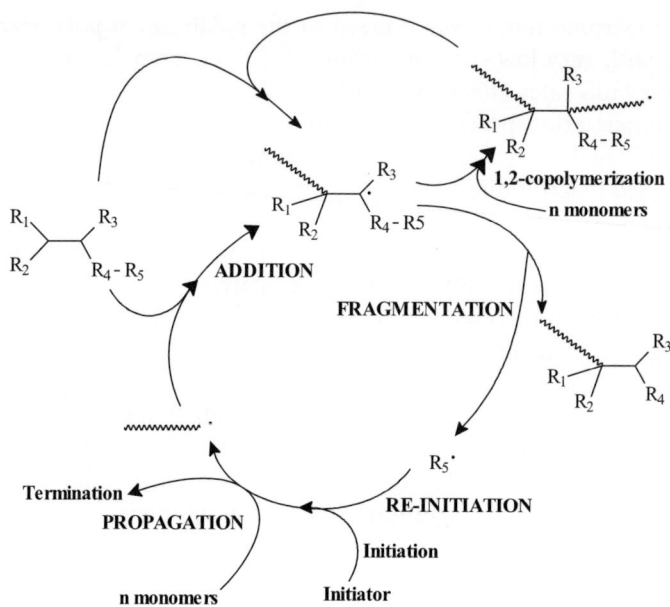

Scheme 58 Free-radical addition–fragmentation processes

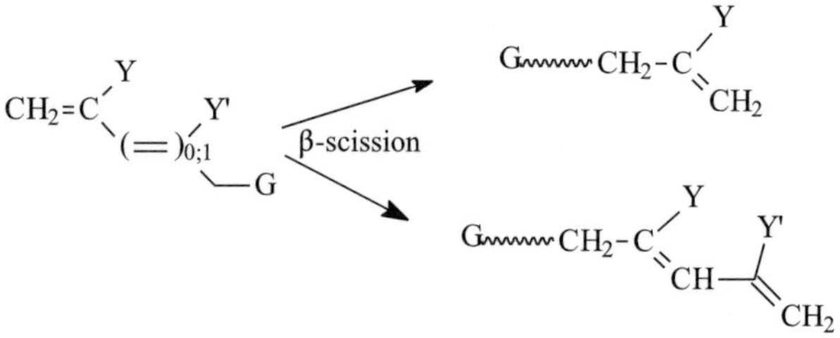

Scheme 59 Synthesis of macromonomers through addition–fragmentation processes

3.3.2
Catalytic Chain Transfer Process

The synthesis of CTAs is often very complex. It involves many reaction steps, which certainly limits the use of the addition–fragmentation process (Scheme 61). CCT may provide an answer to this drawback.

The recent investigations concerning CCT were mainly focused on improving the catalytic system by using new cobalt complexes [302–307]. But CCT reactions usually lead to the synthesis of a large variety of structured monofunctional macromonomers terminated by a vinylic functionality [308, 309]. It is important to note that CCT can be conducted with rate constants C_{tr} hundreds or thousands of times faster than the best mercaptans [69], which

Table 25 CTAs involved in addition–fragmentation leading to a macromonomer structure [54]

Entry	x	Y	G	Y'	Refs.
1	0	CO_2Et	StBu	–	[60, 198]
2	0	CO_2Me	Br	–	[292]
3	0	CO_2H	SCH_2CO_2H	–	[60]
4	0	CN	StBu	/	[407]
5	0	CO_2Ph	$C(Me)_2Ph$	–	[408]
6	1	H	Br	H	[359]
7	1	CH_3	StBu	CO_2Me	[359]

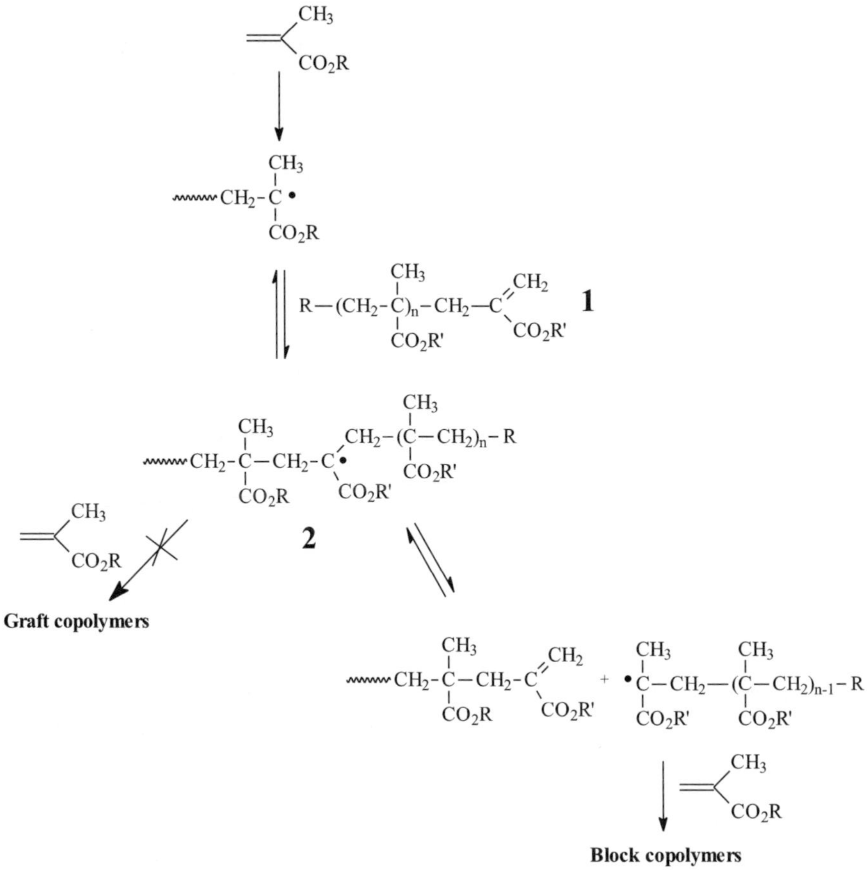

Scheme 60 Macromonomer involved in the synthesis of block copolymers

$$R_n + LCo \xrightarrow{k_{tr}} P_n^= + LCoH$$

$$LCoH + M \xrightarrow{k_r} LCo + R_1$$

Scheme 61 General scheme of catalytic chain transfer

opens up the synthesis of new CTAs for addition–fragmentation [310]. The efficiency of CCT for making good CTAs was mainly proved for MMA [296]. Indeed, macromonomers with $x = 1, 2, 3$ (Scheme 62) were prepared and isolated as pure compounds [311].

The synthesis of these new CTAs was then applied to different methacrylates, leading to macromonomers of PMMA with several functionalities (Scheme 63).

Compound 1 in Scheme 63 [73], which is the trimer of MMA, was directly obtained through CCT of MMA and engaged, after purification, in an addition–fragmentation polymerization of MMA. The transfer constant of such a CTA was about 0.2. But more interesting are compounds 2 and 4 in Scheme 63. Compound 2 in Scheme 63 was obtained by a selective hydrolysis of the MMA trimer, whereas compound 4 in Scheme 63 was obtained through esterification of the trimer of MAA (Scheme 63, 3) previously obtained through nonselective hydrolysis of 1 in Scheme 63. After purification, compounds 2 and 4 in Scheme 63 were used as CTAs in addition-fragmentation of MMA [73]. The calculated transfer constants were 0.3 and 0.16 for 2 and 4 in Scheme 63, respectively. Also oligomers of MMA were obtained, being either α,ω-carboxy from compound 2 in Scheme 63 or α,ω-hydroxy from compound 4 in Scheme 63.

Recently, by using a cobalt complex with porphyrin, CCT led to a methacrylic acid macromonomer in water [312, 313]. The use of this cobalt intermediate added to a continual reinitiation (V501 is the initiator) involves a living character, unlike conventional CCT. The transfer constant at 69 °C was evaluated to be about 4000, which is unexpectedly very high [314].

Addition–fragmentation and CCT are of great interest for the synthesis of macromonomers. Indeed, unlike other radical techniques, they lead to macromonomers in a one-step reaction by directly introducing the chain-end double bond. This double bond is very reactive because it is activated by

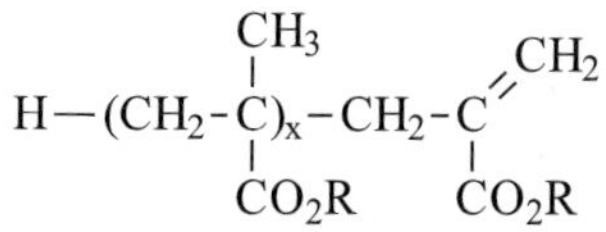

$$H-(CH_2-\underset{\underset{CO_2R}{|}}{\overset{\overset{CH_3}{|}}{C}})_x-CH_2-\underset{\underset{CO_2R}{|}}{C}\overset{CH_2}{\diagup\!\!\!\!\diagdown}$$

Scheme 62 Polymethacrylate macromonomers obtained by catalytic chain transfer ($x = 1, 2, 3$)

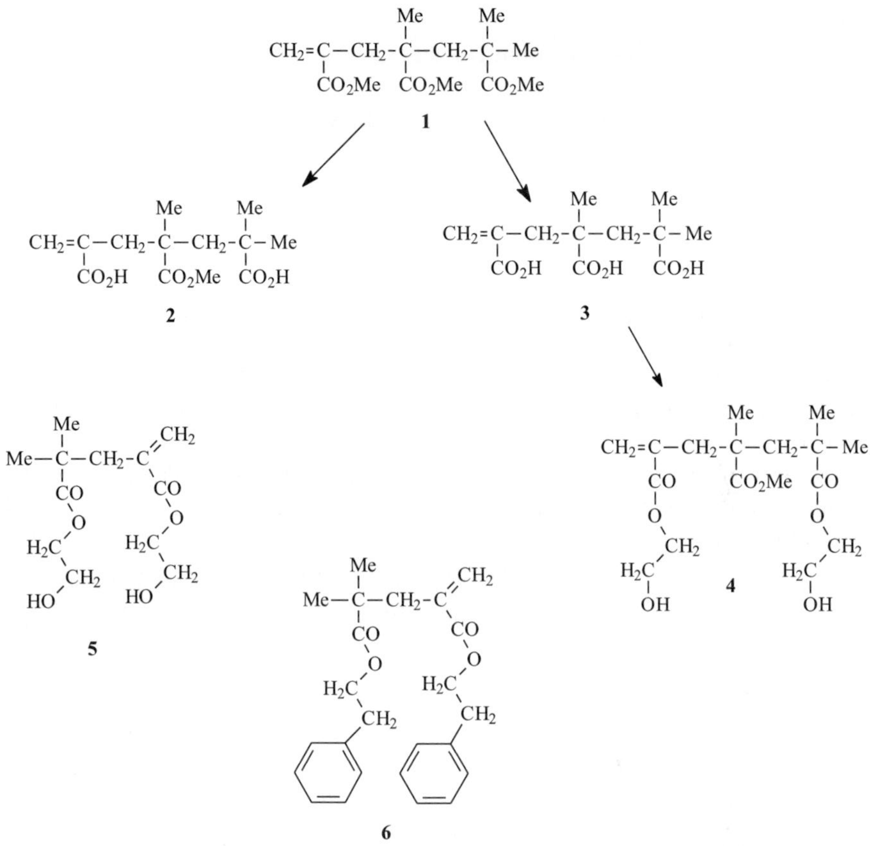

Scheme 63 Several CTAs synthesized by catalytic chain transfer

well-known functions in polymerization. But the main drawback remains the synthesis of CTAs and the high price of the catalytic cobalt complex. In some cases a β-elimination may also occur, leading to block copolymers instead of graft copolymers.

Finally, Chiefari et al. [315–317] suggested another technique leading to the synthesis of addition–fragmentation-type macromonomers but without the use of any CTA. This method, clean, easy, and economical, involves heating a mixture of acrylate or styrene monomer in an appropriate solvent with an azo or peroxy initiator. High temperatures (typically up to 150 °C) are required. To prove the expected mechanism, the authors studied the poly(alkyl acrylate) reactions in the presence (or not) of monomers and by using different conditions. They showed the reaction does not occur without the monomer. Moreover, an increase of the temperature leads to a better yield and a decrease of the molar mass. Macromonomers have been synthesized by this technique with M_n between 10^3 and 10^4 g mol^{-1}.

3.4
Macromonomers Obtained by Atom Transfer Radical Polymerization

Like the telomerization process, ATRP enables the synthesis of two different types of macromonomers: either with a polymerizable double bond or with a polycondensable group. As depicted in Sect. 3.1 on the design of the macromonomers, the polycondensable groups also comprise groups that afford ring-opening polymerization.

3.4.1
Synthesis of Macromonomers with a Polymerizable Double Bond

According to Scheme 64, the resulting oligomers bear an R group (provided by the initiator) in the α position and a halogen atom (provided by the initiator) in the ω position. The macromonomers can also be obtained by ATRP, using three different concepts (Scheme 64):

1. The R group stands for a double bond.
2. The R group may be chemically modified to achieve the double bond.
3. The halogen atom may be chemically modified to achieve the double bond.

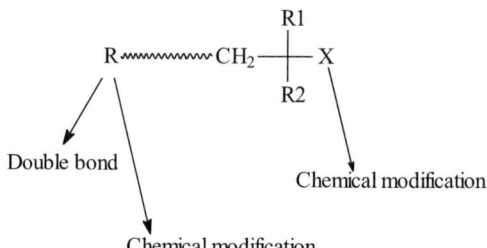

Scheme 64 Different routes to obtain macromonomers with a polymerizable double bond

3.4.1.1
Initiators with an Unsaturated Group

Matyjaszewski et al. first used initiators bearing an unsaturated group for the ATRP process of styrene. In 1998, they [318] used vinyl chloroacetate as the initiator for the ATRP of styrene. As VAc was unreactive towards styrene in radical copolymerization, vinyl chloroacetate was able to initiate the ATRP of styrene (Scheme 65). The resulting PS macromonomers, with molar mass ranging from 5×10^3 to 15×10^3 g mol^{-1}, were copolymerized with N-vinylpyrrolidinone. The amphiphilic copolymers obtained were used as hydrogels.

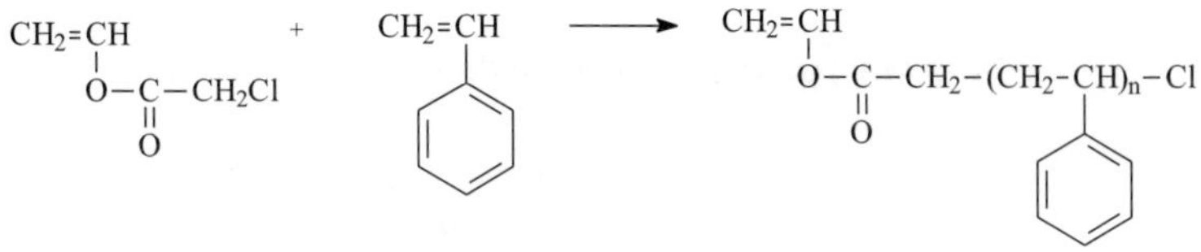

$$CH_2=CH \quad + \quad CH_2=CH \quad \longrightarrow \quad CH_2=CH$$
$$\underset{O}{\overset{|}{O}}-\underset{\|}{C}-CH_2Cl \qquad \qquad \qquad \qquad \underset{O}{\overset{|}{O}}-\underset{\|}{C}-CH_2-(CH_2-CH)_n-Cl$$

Scheme 65 Synthesis of polystyrene macromonomer

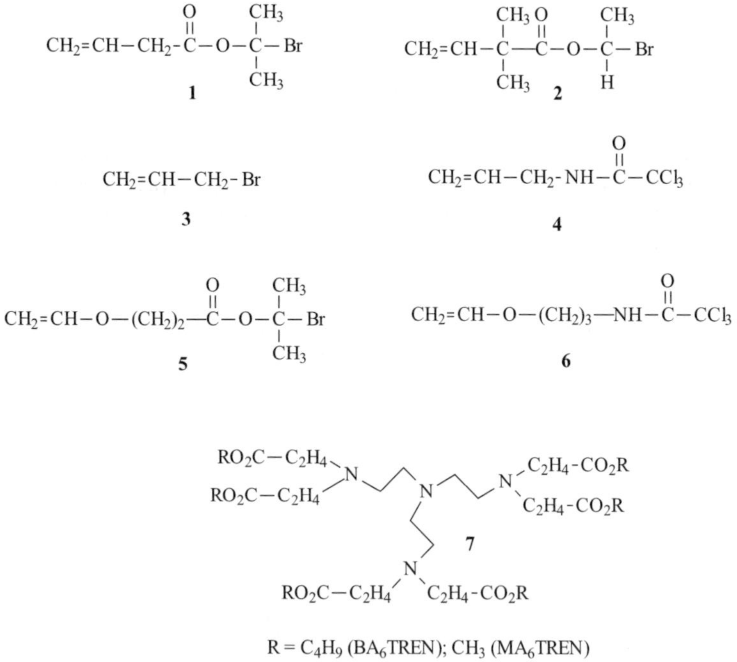

Compound **1**: $CH_2=CH-CH_2-\underset{O}{\overset{\|}{C}}-O-\underset{CH_3}{\overset{CH_3}{\underset{|}{C}}}-Br$

Compound **2**: $CH_2=CH-\underset{CH_3}{\overset{CH_3}{\underset{|}{C}}}-\underset{O}{\overset{\|}{C}}-O-\underset{H}{\overset{CH_3}{\underset{|}{C}}}-Br$

Compound **3**: $CH_2=CH-CH_2-Br$

Compound **4**: $CH_2=CH-CH_2-NH-\underset{O}{\overset{\|}{C}}-CCl_3$

Compound **5**: $CH_2=CH-O-(CH_2)_2-\underset{O}{\overset{\|}{C}}-O-\underset{CH_3}{\overset{CH_3}{\underset{|}{C}}}-Br$

Compound **6**: $CH_2=CH-O-(CH_2)_3-NH-\underset{O}{\overset{\|}{C}}-CCl_3$

Compound **7**: tris(2-aminoethyl)amine derivative with $RO_2C-C_2H_4$ and $C_2H_4-CO_2R$ arms on each nitrogen.

$R = C_4H_9$ (BA$_6$TREN); CH_3 (MA$_6$TREN)

Scheme 66 Different allylic and vinylic initiators

Zeng et al. [319] extended the use of unsaturated initiator to allyl-type and vinyl-type initiators (Scheme 66).

Several ligands were used with allyl-type and vinyl-type initiators, such as 1,1,4,7,10,10-hexamethyltriethylenetetramine (HMDETA), N,N,N',N',N'-pentamethyldiethylenetriamine (PMDETA), or compound **7** in Scheme 66. Zeng et al. showed that the combination of initiator **1** in Scheme 66 with BA$_6$TREN or initiator **4** in Scheme 66 with BA$_6$TREN gave the best control of the molar mass for the ATRP of 2-(dimethylamino)ethyl methacrylate. These allylic macromonomers are then able to copolymerize with acrylamide.

Concerning the vinyl-ether-type macromonomers (obtained with compounds **5** and **6** in Scheme 66), their copolymerization was studied with several monomers. The authors observed that the copolymerization was not efficient with styrene or methacrylates, unlike acrylates. However, with acrylates it is necessary to stop the reaction before completion to keep intact the unsaturated group at the chain end.

3.4.1.2
Modification of the Functional Group Provided by the Initiator

To obtain the unsaturation, the methodology used in the telomerization process with a monofunctional transfer agent can be extrapolated to the ATRP process. However, unlike telomerization, it is necessary to eliminate the chain-end halogen atom to avoid any side reactions. Several techniques may overcome this problem. For instance, Neugebauer et al. [320] suggested the grafting-from method consisting of several steps, according to Scheme 67.

Moreover, Schoen et al. [321] employed a new strategy to eliminate the chain-end halogen atom, based on a transfer reaction onto the ligand. They investigated the ATRP of acrylate monomer in the presence of 2-hydroxy-

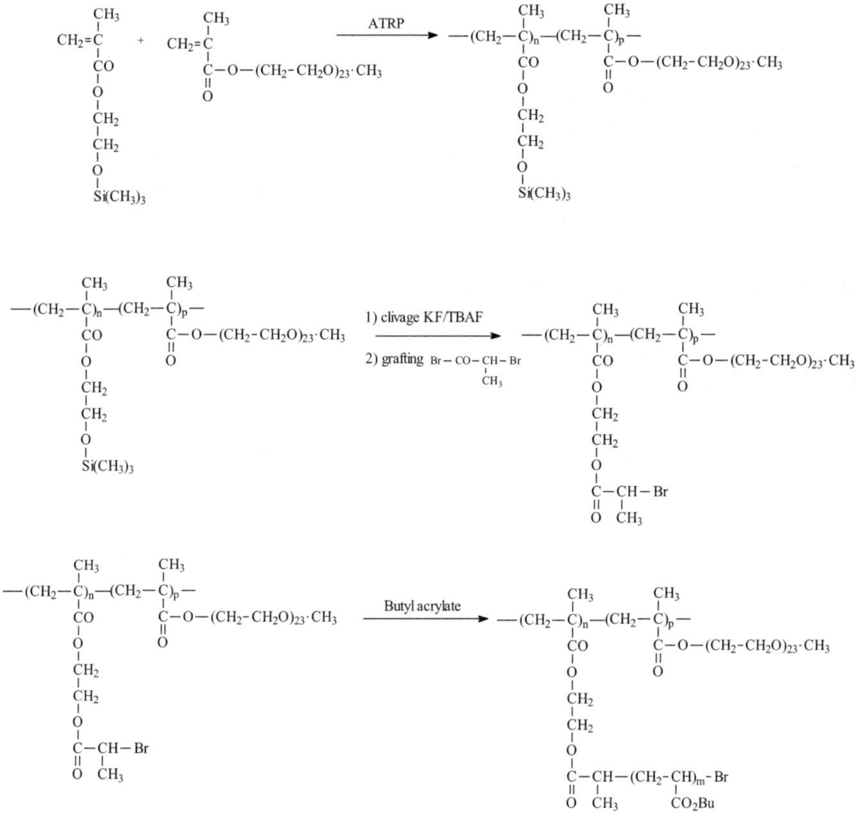

Scheme 67 Elimination of halogen atom by grafting-from

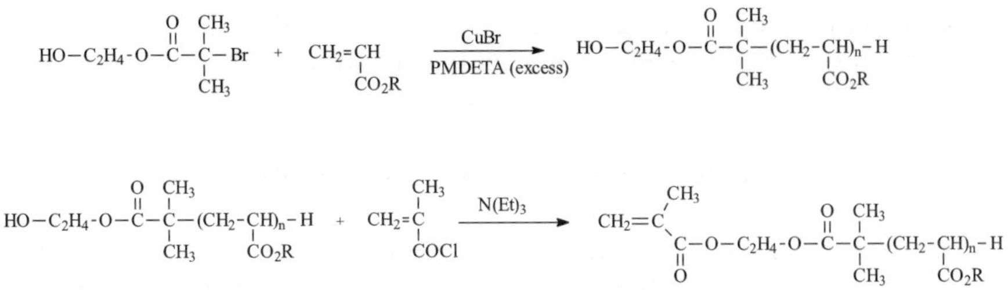

Scheme 68 Synthesis of poly(alkyl acrylate) macromonomer

ethyl-2-bromoisobutyrate as a functionalized initiator and also with a large excess of PMDETA ligand (relative to CuBr). These special conditions allow a hydrogen transfer from the ligand onto the ω position (Scheme 68). The unsaturation is then obtained by reaction of the terminal group with methacryloyl chloride.

In a previous work, Cheng et al. [322] performed the same synthesis but without any ligand excess. The resulting macromonomer was similar to that described in Scheme 69 but with a chain-end bromine atom. This macromonomer was polymerized by the ATRP process, leading to a hypergrafted polymer. The Mark–Houwink coefficient was 0.47, which characterized the hyperbranched structure. Hydrolysis of such a polymer led to the corresponding poly(acrylic acid). Similarly, Hua et al. [323] performed the synthesis of brushlike poly(acrylic acid).

3.4.1.3
Modification of the Chain-End Halogen Atom

Muehlebach [324] developed an original method that consists of replacing the terminal bromine atom by a methacrylate function (Scheme 69).

The rate of methacrylate functionalization is above 90% for molar masses ranging between 1500 and 24 000 g mol^{-1}. The efficiency of such functionalization was evidenced by further copolymerization with 2-(dimethylamino)-ethyl methacrylate.

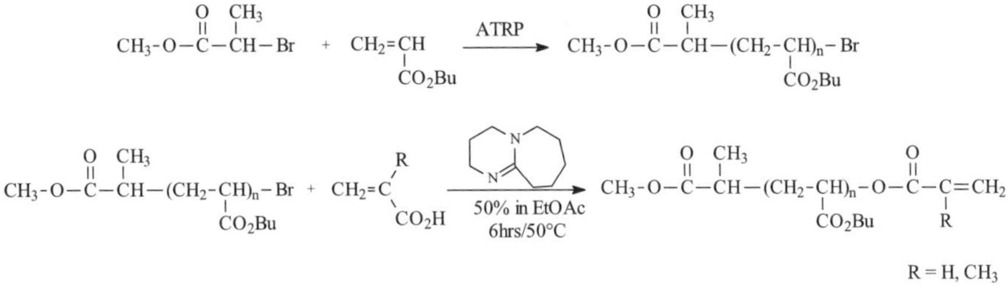

Scheme 69 Synthesis of poly(*n*-butyl acrylate) macromonomer

In a similar study, Schulze et al. [325] suggested the synthesis of polypropylene-g-PS copolymers by copolymerization of PS macromonomer with propylene, using metallocene catalysis. They first synthesized by ATRP oligomers of styrene with a chain-end bromine or chlorine atom. After purification, macromonomers are obtained by reaction of oligomers with allyl trimethylsilane [326, 327], followed by addition with a Lewis acid (TiCl$_4$) without any monomer (Scheme 69). In these conditions, the terminal halogen is replaced by a carbocation and Ti$_2$Cl$_9^-$. The carbocation obtained will directly lead to the allylic double bond. ^{1}H NMR easily characterized the absence of the terminal halogen [328]. The macromonomers synthesized exhibited molar masses ranging from 1200 to 18 300 g mol^{-1} with PDI close to 1.2. The macromonomer functionality is almost 1 and corresponds to that of the initial halogen [156].

Recently, Couvreur et al. [329, 330] proposed the synthesis of acrylic-type macromonomers by direct substitution of the terminal halogen atom. The nucleophilic modification of the bromide end group of both types of polymers to a polymerizable acrylate or methacrylate group has been achieved in order to obtain a wide range of macromonomers. Such macromonomers are widely used as starting products for graft copolymers and other highly branched polymer architectures. The success of this end-group modification has been studied in detail by MALDI-TOF MS and NMR. Then, ATRP of these macromonomers was successfully performed in order to synthesize well-defined comblike poly(macromonomers) with controlled chain length and low polydispersity.

In a similar way, Norman et al. [331] synthesized PMMA oligomers by ATRP. The methacrylic double bond of the resulting macromonomer was directly obtained by elimination of the terminal halogen by catalytic CTAs, such as 5,10,15,20-tetraphenyl-21H,23H-porphine cobalt(II) [Co(tpp)]

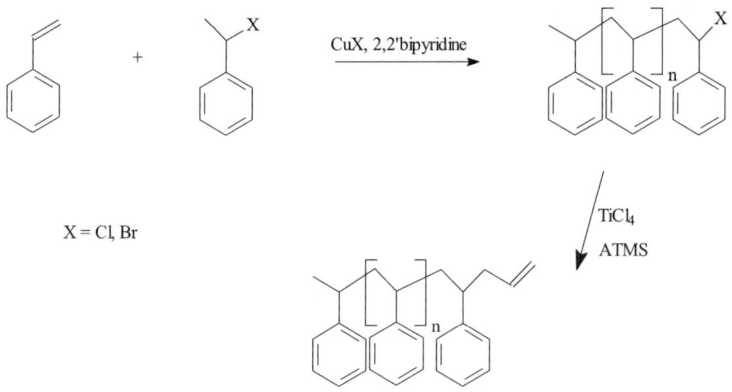

Scheme 70 Functionalization of oligomers obtained by ATRP (addition of allyl trimethylsilane, *ATMS*)

and bis[(2,3-butanedione dioximato) (2-) $O:O'$] tetrafluorodiborato (2-) N,N',N'',N''' cobalt(II), near the end of an atom-transfer polymerization. Low molecular weight, narrow polydispersity PMMA polymers prepared by ATRP have been converted in high yield (85%) to the ω-unsaturated PMMA species by the in situ addition of Co(tpp) to an ATRP reaction mixture. These species have been copolymerized successfully with ethyl acrylate using an azo initiator with little or no sign of the original macromonomer. Narrow polydispersity diblock copolymers (of MMA and BMA) prepared by ATRP have also been converted to the corresponding unsaturated end-group species by the addition of Co(tpp) in solution to a "live" reaction mixture.

3.4.2
Synthesis of Macromonomers with Polycondensable Groups

In the last decade, several new reactive groups in polycondensation have been employed as ATRP initiators, e.g., lactones [332, 333], as follows:

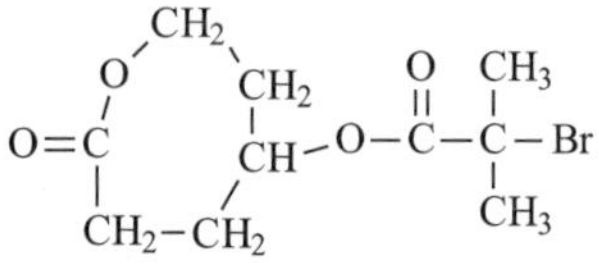

These new compounds serve both as a monomer in living ring-opening polymerization and as an initiator in the ATRP process. For instance, the resulting PMMA macromonomers, bearing PMMA grafting groups, were further copolymerized with ε-caprolactone via ring-opening polymerization to form graft polyester copolymers.

Another macromonomer, bearing a pyrrole end group with a methacrylate lateral chain, was synthesized and copolymerized by the same authors [334]. The macromonomer was synthesized in the presence of initiator **8**, according to Scheme 71:

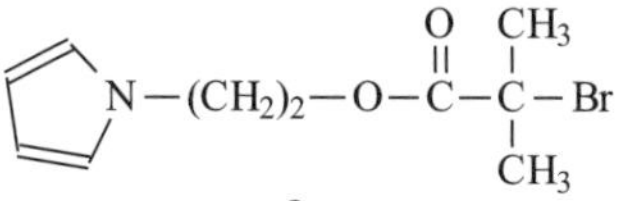

Scheme 71 Pyrrole- or oxazoline-terminated bromine initiators in the ATRP process

This macromonomer may be employed in the field of electrical conductive polymers. Methacrylate chains provide better processing properties than the usual (pyrole) polymers.

Interestingly, we can mention the oxazoline-terminated PS macromonomers [335], obtained by ATRP in the presence of initiators **9** and **10** in Scheme 71.

Furthermore, Cianga and Yagci [336] performed the synthesis of graft copolymers in which the lateral chains were obtained through an ATRP process. The lateral chains were good organophilic compounds (such as PS) in order to increase the organic solubility of the main chain, i.e., polyphenylene, for the graft copolymer. The lateral chain can be obtained with the corresponding dibromine initiator:

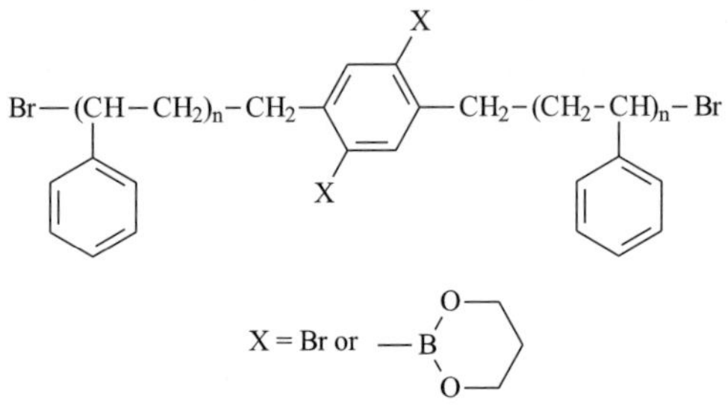

$$Br-(CH-CH_2)_n-CH_2-\text{Ar}(X)(X)-CH_2-(CH_2-CH)_n-Br$$

$$X = Br \text{ or } -B\begin{pmatrix}O-\\O-\end{pmatrix}$$

The graft copolymers were then obtained either by Suzuki [with Pd(PPh$_3$)$_4$] or Yamamoto (with NiCl$_2$) reaction [337]:

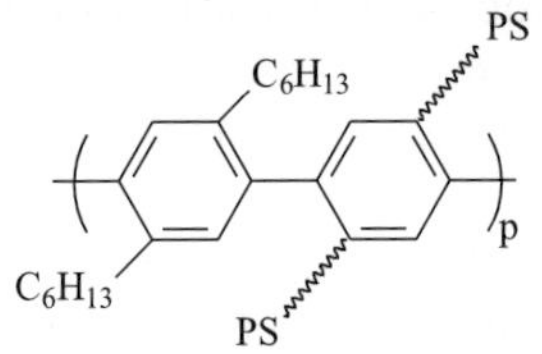

Furthermore, the group of Deimede [338–340] performed the synthesis of α-dicarboxy end-functionalized PS macromonomers by using ATRP (Scheme 72). Further polycondensation with dihydroxy end functionalized poly(ethylene oxide) led to alternating branched PS/poly(ethylene oxide) poly(macromonomers) (Scheme 72). These novel amphiphilic compounds afforded the formation of stable micelles, especially in THF or dioxane.

In the course of synthesizing new macromonomers, ATRP is one of the most appropriate radical techniques, witnessed by the high number of publications in this area. First, like for other LRP, the macromonomers obtained possess macromolecular chains with very low PDI, which confer specific properties. Second, there are several possibilities for synthesizing macromonomers through the ATRP process. Indeed, macromonomers can be

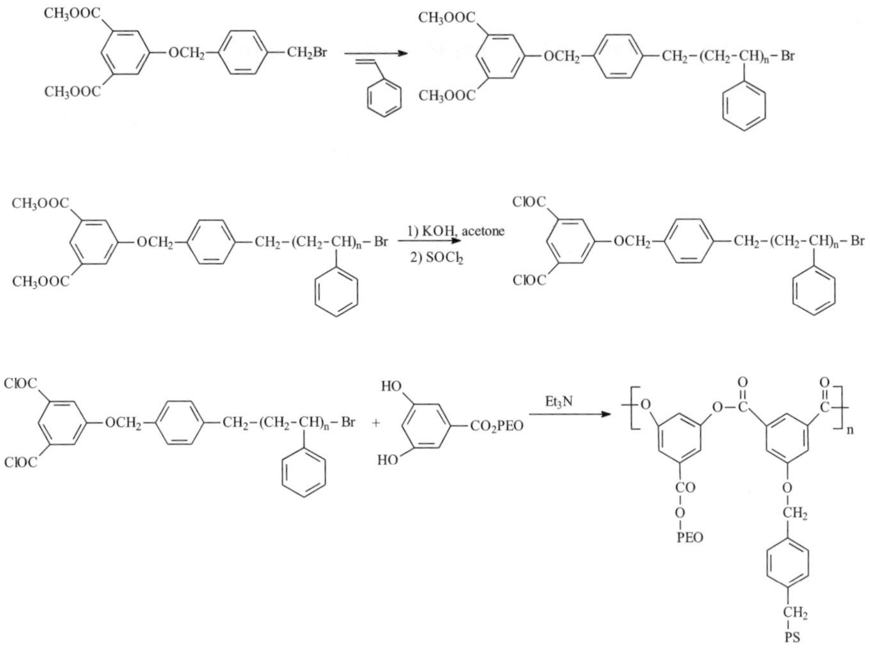

Scheme 72 Synthesis of alternating branched polystyrene/poly(ethylene oxide) poly(macromonomers)

obtained by simply using an initiator with a reactive double bond but also by chemical modification of both extremities, i.e., the terminal halogen or the functional group provided by the initiator. The first strategy, based on the use of an unsaturated group carried by the initiator, is less developed because it implies the use of unreactive double bonds during the ATRP process. Despite this limitation, the number of initiators with unsaturated groups used in ATRP remains high. The second strategy, i.e., chemical modification of extremities, gives most of the macromonomers by ATRP. Indeed, the methods for replacing the halogen atom by a reactive double bond are common and easily reproducible, e.g., reaction with allyl trimethylsilane in the presence of TiCl$_4$. Hence, the syntheses of macromonomers by the ATRP process have certainly not been totally explored. Several synthetic strategies can be used to achieve the targeted macromonomer structure, but also most vinyl monomers are efficient in ATRP. However, elimination of HX (where X is a halogen atom) often occurs, leading to a chain-end double bond which does not enable the synthesis of macromonomers.

3.5
Macromonomers Obtained by Nitroxide-Mediated Polymerization

As already described in Scheme 35 (Sect. 2.7), the macromolecules obtained by the NMP process exhibit the general following structure: a functional group provided by the initiator at a chain end (α position) and an aminoxyl function at the other chain end (ω position). A macromonomer structure may be achieved by modifying one of the two positions.

3.5.1
Modification of the ω Position

Kuckling and Wohlrab [341] polymerized 2-vinylpyridine in the presence of hydroxy-TEMPO (Scheme 73). The macromonomer was obtained by reacting the hydroxyl group, situated at the ω position, with acryloyl chloride.

The poly(vinylpyridine) chloride was then copolymerized with NIPAM in the presence of N,N'-methylenebisacrylamide to obtain graft copolymer gels. These gels were found to be temperature- and pH-dependent. But above 33 °C, the authors showed aggregation of the poly(NIPAM) phase and a pH > 5.5 leads to aggregation of the poly(vinylpyridine). However, the pH effect remains minor compared with that of temperature.

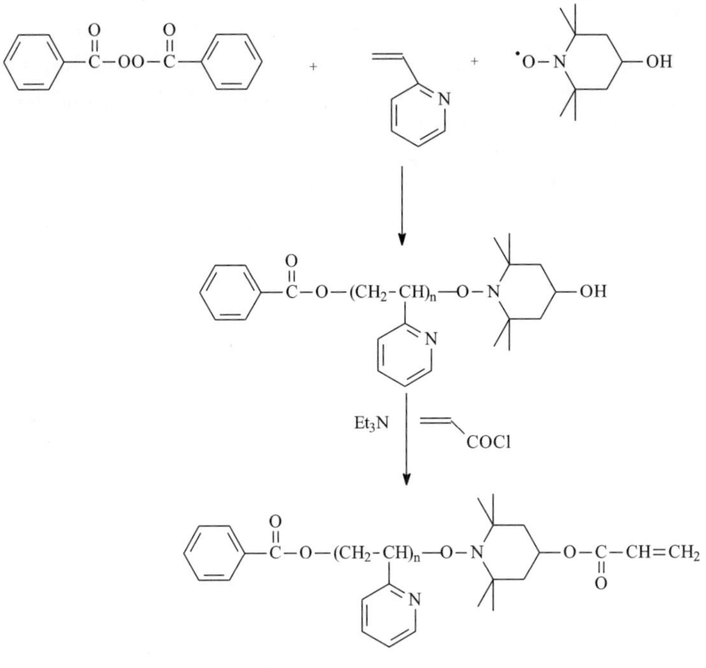

Scheme 73 Synthesis of poly(2-vinylpyridine) macromonomer

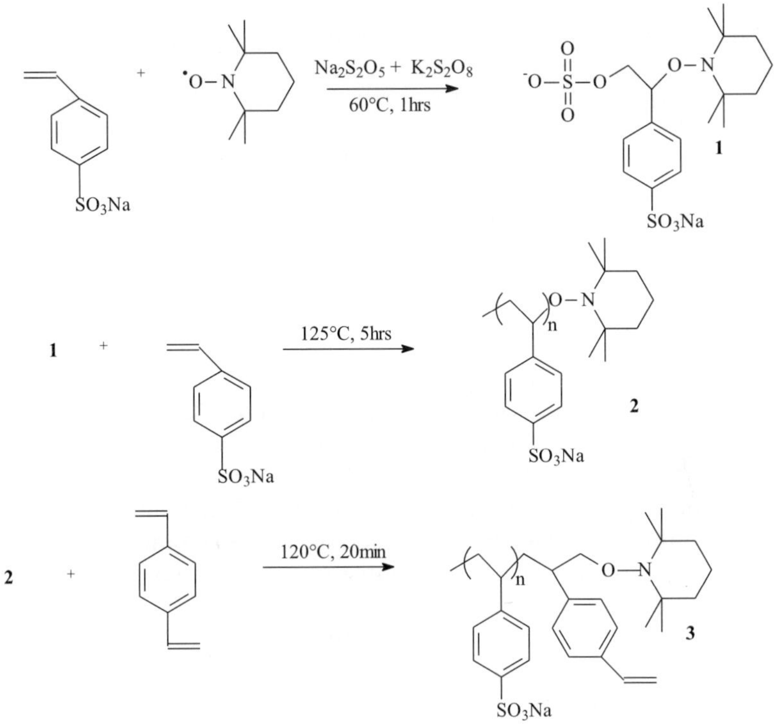

Scheme 74 Synthesis of poly(sodium styrenesulfonate) macromonomer

Other teams worked on the functionalization of the aminoxyl group situated at the ω position. For instance, the method of Ding et al. [342] is original for the synthesis of a novel series of poly(sodium styrenesulfonate) (PSSNa) macromonomers (compound 3 in Scheme 74) based on stable free radical polymerization in the presence of TEMPO.

The (PSSNa) macromonomer was then copolymerized with styrene by emulsion polymerization to yield proton exchange membranes with sodium ions. The original structure of these graft copolymers (i.e., hydrophilic part owing to PSSNa) affords good ionic conductivity and may become a good model of NAFION membranes.

3.5.2
Modification of the α Position

To obtain a macromonomer starting from the α position consists of a chemical modification on the function provided by the initiator used during the NMP process. For instance, Hawker et al. [343] replaced the benzoyl group, provided by the benzoyl peroxide initiator, by a hydroxyl group. This latter

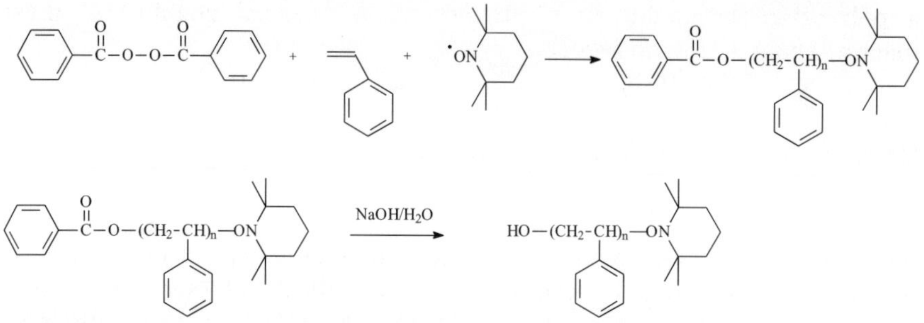

Scheme 75 Chemical modification of the α position

group is then able to react with acryloyl chloride to produce the reactive double bond of the macromonomer (Scheme 75).

Similarly, Hawker et al. [220, 344] synthesized original PS macromonomers with two amine groups situated at the α position. These amine groups can further react through a condensation reaction. This macromonomer was synthesized in the presence of a peculiar diazoic initiator 4 prepared in a first step, according to Scheme 76.

NMP leads to a macromonomer structure by modification of either the α position or the ω position, i.e., the aminoxyl function. The α position is brought by the intiator and a chemical modification usually leads to macromonomers with two polycondensable groups. Moreover, the functionalization of the aminoxyl function will lead to macromonomers with

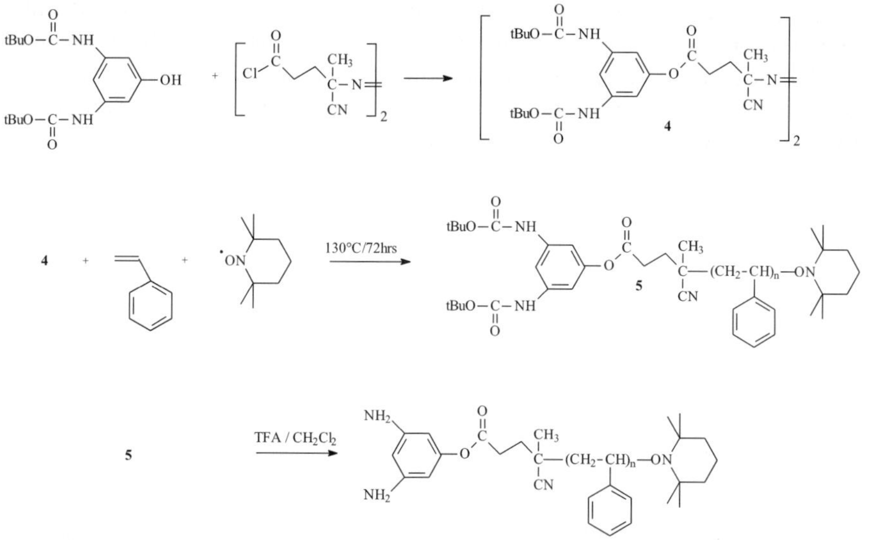

Scheme 76 Synthesis of polystyrene macromonomer with two condensable amine groups

a polymerizable double bond. However, the thermal stability of these macromonomers remains weak.

3.6
Macromonomers Obtained by Other Techniques

Among all the radical processes previously described, more specific techniques can lead to the synthesis of macromonomers. For instance, the use of borans in radical polymerization, or the radical polymerization based on unimolecular terminations, may allow macromonomers to be obtained. These specific techniques will be briefly summarized as only a few workers have investigated the use of such techniques, aiming at the synthesis of macromonomers.

Chung [345] is certainly one of the best specialists in the use of borans in radical polymerization. This original method to obtain macromonomers is described in Scheme 77.

A plot of the molar mass versus the monomer conversion produces a straight line that characterizes a living process. The molar masses ranged between 10^4 and 10^5 g mol^{-1} and the macromonomer structure was perfectly established by ^{13}C NMR.

Some Japanese teams developed a novel technique, based on unimolecular termination, which allows separating both initiation and termination (or transfer) processes. After growing chains are obtained, the macroradical formed is able to react with another molecule (mainly unsaturated) to lead to a stable radical. This one may transfer to give another radical able to reinitiate a polymerization. This process was developed, aiming at synthesizing either telechelic oligomers [85] (Scheme 78) or macromonomers [86] (Scheme 79).

Concerning the synthesis of VAc macromonomer, Fukutomi et al. [270] showed a functionality of 1.78 per polymer chain. This result was attributed to side reactions of chloromethylstyrene (end-capping agent) onto $-N=$ sites

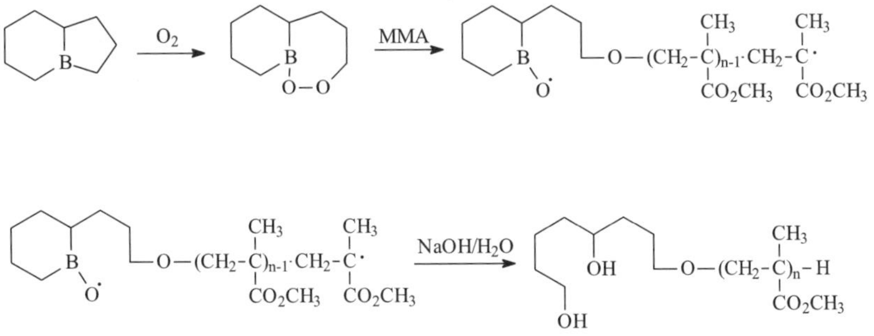

Scheme 77 Synthesis of PMMA macromonomer with boran

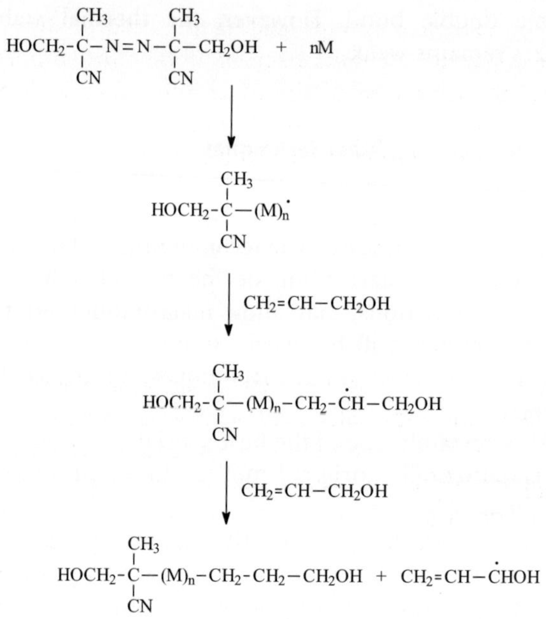

Scheme 78 Synthesis of hydroxyl-telechelic oligomers by using an iniferter system [85]

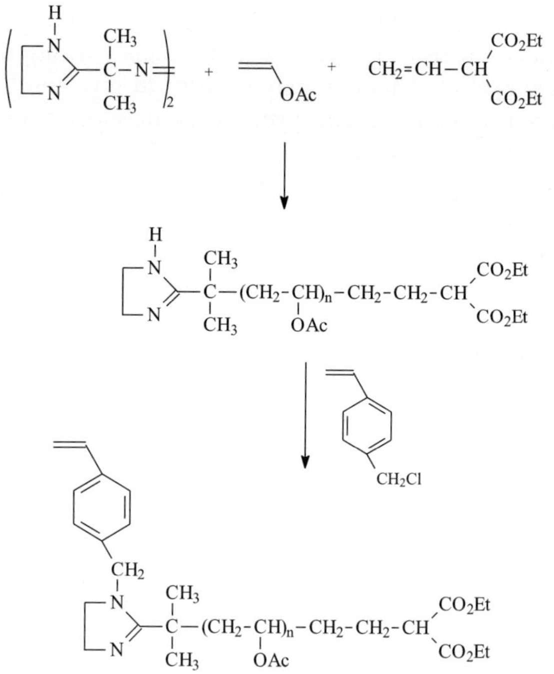

Scheme 79 Synthesis of poly(vinyl acetate) macromonomer by using an iniferter system

of the terminal imidazoline group, leading to some quaternization. After hydrolysis of the poly(vinyl alcohol) macromonomer, the authors investigated the emulsion copolymerization of the macromonomer with MMA, resulting in microspheres in a water–alcohol solution. Indeed, the hydrophilic macromonomer afforded stabilization of the emulsion during the copolymerization.

4
Conclusion

This review aimed at providing the new designs of macromonomers and telechelic oligomers and especially their syntheses using both conventional polymerization and CRP. Concerning conventional radical polymerizations, this review supplied different studies since the reviews of Boutevin [2] and Rempp and Franta [3]. However, the synthesis of macromonomer and telechelic structures by using conventional radical polymerizations has not been described in detail in this review. Indeed, unlike the conventional radical techniques, the controlled radical ones represent a major breakthrough for the syntheses of macromolecular structures because they afford very good control of the macromolecular architectures (control of the molar masses and low PDI). Hence, this review has shown how ATRP, NMP, addition-fragmentation processes, and ITP can lead to both macromonomers and telechelic oligomers. For all these living techniques, the oligomers obtained bear a reactive function at the chain end, e.g., xanthate, bromine, iodine, or aminoxyle. The synthesis of telechelic oligomers or macromonomers requires a chemical modification of these reactive functions. The literature offers many possibilities to modify such reactive groups: radical reactions, nucleophilic substitution, etc.

To the authors' knowledge, most telechelic oligomers and macromonomers are obtained by ATRP. This may be explained by a relatively easy replacement of the terminal halogen atom. However, even after chemical modification, CuBr traces remain in the final product, which represents a major drawback for further industrial developments.

The syntheses of macromonomers and telechelic oligomers by using LRP have not been developed industrially yet. Indeed, unlike conventional radical polymerizations (i.e., telomerization and DEP), the cost of CRPs still remains very high. However, despite such high cost, new possibilities are now opened up for telechelic oligomers and macromonomers obtained by CRPs. This concerns, for instance, the recent investigations into the nanostructuring and especially through noncovalent linkages. The work of Lohmeijer et al. [224] illustrated the synthesis of "metallo-supramolecular copolymers," and that of Leibler [346] linked bifunctional and trifunctional oligomers by hydrogen bonding. For now, the innovative works mainly concern (macro)molecules of low molecular weight, obtained by polycondensation or radical copolymerization, in which the linking groups are statistically dispersed into the chain. But these new telechelic oligomers obtained by LRP may help in building more complex macromolecular structures. On the other hand, the synthesis of new macromonomers aims at obtaining new graft copolymers with controlled architectures. These new types of graft copolymers should provide interesting properties and should find new applications in various areas, such as biological applications (transfer of peptides/proteins) or new membranes for fuel cells.

In conclusion, this review has considered the whole range of synthetic methodologies based on radical polymerizations to achieve the desired telechelic or macromonomer structures. It is very difficult to rank the various synthetic techniques (radical polymerizations and chemical modifications), especially in terms of functionality. Indeed, the polymerization process is well adapted to a monomer in certain conditions, but may not be reproduced in different conditions. Many side reactions may also occur, leading to dramatic loss of the functionality. However, this review shows how to get the functionality as close as 1 or 2 (depending on the structure), especially by moving from conventional radical polymerizations to CRPs. For instance, hydroxy-telechelic polybutadiene was first obtained by a current commercial process using hydrogen peroxide as the initiator in DEP conditions. But, owing to side reactions (grafting sites), the hydroxy functionality was about 2.3 for $M_n = 1000$ g mol^{-1} and even 2.7 for $M_n = 2500$ g mol^{-1}. The functionality was then decreased by polymerizing butadiene by CRP. Indeed, the NMP of butadiene in the presence of TEMPO followed by a continuous elemination of TEMPO units by sublimation led to an average functionality of 2.06 [217, 219]. This example illustrates the difficulty of exactly matching the bifunctionality and also clearly demonstrates the utility of CRP. The success of functionalization by using CRP may be explained by the low number of termination reactions owing to the stability of propagating radical intermediate. Taking into account such an assumption, anionic or cationic polymerizations would certainly allow a functionality of 1 or 2 to be achieved. For instance, hydroxy-telechelic–polybutadiene was recently synthesized by Schwindeman et al . [347]. But such techniques still require drastic conditions.

References

1. Bielawski CW, Jethmalani JM, Grubbs RH (2003) Polymer 44:3721–3726
2. Boutevin B (1990) Adv Polym Sci 94:69–105
3. Rempp P, Franta E (1984) Adv Polym Sci 58:1
4. Ito K, Kawaguchi S (1999) Adv Polym Sci 142:129–178
5. Ito K (1998) Prog Polym Sci 23:581–620
6. Ito K (1997) Kobunshi 46:741–742

7. Hawker CJ, Bosman AW, Harth E (2001) Chem Rev 101:3661–3688
8. Matyjaszewski K, Xia J (2001) Chem Rev 101:2921–2990
9. Rizzardo E, Chiefari J, Mayadunne R, Moad G, Thang S (2001) Macromol Symp 174:209–212
10. Gaynor SG, Wang J-S, Matyjaszewski K (1995) Macromolecules 28:8051–8056
11. Gaynor SG, Wang JS, Matyjaszewski K (1995) Polym Prepr Am Chem Soc Div Polym Chem 36:467–468
12. Mayo FR (1943) J Am Chem Soc 65:2324–2329
13. Boutevin B (2000) J Polym Sci Part A Polym Chem 38:3235–3243
14. Bellesia F, Forti L, Gallini E, Ghelfi F, Libertini E, Pagnoni UM (1998) Tetrahedron 54(27):7849–7856
15. Bellesia F, Forti L, Ghelfi F, Pagnoni UM (1997) Synth Commun 27(6):961–971
16. Ameduri B, Berrada K, Boutevin B, Bowden RD, Pereira L (1991) Polym Bull 26:377–382
17. Ameduri B, Berrada K, Boutevin B, Bowden RD (1992) Polym Bull 28:389–394
18. Ameduri B, Berrada K, Boutevin B, Bowden RD (1992) Polym Bull 28:497–503
19. McKierna RL, Cardoen G, Boutevin B, Ameduri B, Gido SP, Penelle J (2003) Macromol Chem Phys 204:961–969
20. Esselborn E, Fock J, Knebelkamp A (1996) Macromol Symp 102:91–98
21. Fock J, Knebelkamp A (1997) Eur Pat Appl
22. Esselborn E, Fock J (1996) Eur Pat Appl 12 pp
23. Esselborn E, Fock J (1994) Eur Pat Appl 18 pp
24. Esselborn E, Fock J (1996) Eur Patent Appl 708 115
25. Esselborn E, Fock J, Knebelkamp A (1996) Macromol Symp 102:91–98
26. Schroder N, Konczol L, Doll W, Mulhaupt R (1998) J Appl Polym Sci 70:785–796
27. Liu P, Ding H, Liu J, Yi X (2002) Eur Polym J 38:1783–1789
28. Polowinski S, Bortnowska-Barela B (1981) J Polym Sci Polym Chem Ed 19:51–55
29. Brosse JC, Derouet D, Epaillard F, Soutif JC, Legeay G, Dusek K (1986) Adv Polym Sci 81:167–223
30. Tobolsky AV (1958) J Am Chem Soc 80:5927–5929
31. Berger KC, Meyerhoff G (1975) Makromol Chem 176:1983–2003
32. Berger KC (1975) Makromol Chem 176:3575–3592
33. Bessiere JM, Boutevin B, Loubet O (1995) Eur Polym J 31:573–580
34. Bamford CH, Jenkins AD (1955) Nature 176:78
35. Ohishi H, Kishimoto S, Nishi T (2000) J Appl Polym Sci 78:953–961
36. Ohishi H, Nishi T (2000) J Polym Sci Part A Polym Chem 38:299–309
37. Ohishi H, Ikehara T, Nishi T (2001) J Appl Polym Sci 80:2347–2360
38. David G, Boutevin B, Robin J-J, Loubat C, Zydowicz N (2002) Polym Int 51:800–807
39. Bickel AF, Waters WA (1950) Recl Trav Chim Pays-Bas Belg 69:1490–1494
40. Beyou E, Chaumont P, Chauvin F, Devaux C, Zydowicz N (1998) Macromolecules 31:6828–6835
41. David G, Robin J-J, Boutevin B (2001) J Polym Sci Part A Polym Chem 39:2740–2750
42. David G, Boutevin B, Robin JJ (2002) J Polym Sci Part A Polym Chem 41:236–247
43. Cypcar CC, Camelio P, Lazzeri V, Mathias LJ, Waegell B (1996) Macromolecules 29:8954–8959
44. Banthia AK, Chaturvedi PN, Jha V, Pendyala VNS (1989) Adv Chem Ser 222:343–358
45. David G, Loubat C, Boutevin B, Robin JJ, Moustrou C (2002) Eur Polym J 39:77–83
46. Boutevin B, Bosc D, Rousseau A (1997) Desk Ref Funct Polym 489–503
47. Alric J, David G, Boutevin B, Rousseau A, Robin J-J (2002) Polym Int 51:140–149

48. Saint-Loup R, Manseri A, Ameduri B, Lebret B, Vignane P (2002) Macromolecules 35:1524–1536
49. Vignane P, Lebret B, Ameduri B, Manseri A, Saint-Loup R (2001) Fr Patent Appl 2810668
50. Smithenry DW, Kang M-S, Gupta VK (2001) Macromolecules 34:8503–8511
51. Colombani D, Beliard I, Chaumont P (1996) J Polym Sci Part A Polym Chem 34:893–902
52. Yamada B, Kobatake S (1994) Prog Polym Sci 19:1089–1131
53. Rizzardo E, Meijs GF, Thang SH (1995) Macromol Symp 98:101–123
54. Colombani D, Chaumont P (1998) Acta Polym 49:225–231
55. Colombani D (1999) Prog Polym Sci 24:425–480
56. Colombani D, Chaumont P (1996) Prog Polym Sci 21:439–503
57. Monteiro MJ, Bussels R, Wilkinson TS (2001) J Polym Sci Part A Polym Chem 39:2813–2820
58. Kochi JK (ed) (1973) Free radicals, vols I–II. Wiley, New York
59. Pryor WA (ed) (1982) Free radicals in biology, vols I–V. Academic, New York
60. Meijs GF, Morton TC, Rizzardo E, Thang SH (1991) Macromolecules 24:3689–3695
61. Meijs GF, Rizzardo E, Thang SH (1988) Macromolecules 21:3122–3124
62. Montaudon E, Rakotomanana F, Maillard B (1985) Tetrahedron 41:2727–2735
63. Colombani D, Chaumont P (1994) J Polym Sci Part A Polym Chem 32:2687–2697
64. Colombani D, Chaumont P (1994) Macromolecules 27:5972–5978
65. Colombani D, Maillard B (1994) J Chem Soci Perkin Trans 2:745–752
66. Enikolopov NS, Korolev GV, Marchenko AP, Ponomarev GV, Smirnov BR, Titov VI (1980) USSR Patent 664434
67. Gridnev AA, Ittel SD (2001) Chem Rev 101:3611–3659
68. Gridnev AA, Simonsick WJ Jr, Ittel SD (2000) J Polym Sci Part A Polym Chem 38:1911–1918
69. Gridnev A (2000) J Polym Sci Part A Polym Chem 38:1753–1766
70. Barner-Kowollik C, Davis TP, Stenzel MH (2004) Polymer 45:7791–7805
71. Haddleton DM, Topping C, Hastings JJ, Suddaby KG (1996) Macromol Chem Phys 197:3027–3042
72. Haddleton DM, Topping C, Kukulj D, Irvine D (1998) Polymer 39:3119–3128
73. Hutson L, Krstina J, Moad CL, Moad G, Morrow GR, Postma A, Rizzardo E, Thang SH (2004) Macromolecules 37:4441–4452
74. Otsu T, Kuriyama A (1984) J Macromol Sci Chem A 21:961–977
75. Otsu T, Matsumoto A (1998) Adv Polym Sci 136:75–137
76. Otsu T (2000) J Polym Sci Part A Polym Chem 38:2121–2136
77. Bledzki A, Braun D, Titzschkau K (1983) Makromol Chem 184:745–754
78. Bledzki A, Braun D (1986) Polym Bull 16:19–26
79. Bledzki A (1987) Pr Nauk Politech Szczecin 337:123
80. Bledzki A, Balard H, Braun D (1988) Makromol Chem 189:2807–2822
81. Bledzki AK (1990) Polimery 35:408–412
82. Braun D, Steinhauer-Beisser S (1997) Eur Polym J 33:7–12
83. Roussel J, Boutevin B (2001) Polym Int 50:1029–1034
84. Nair CPR, Clouet G, Chaumont P (1989) J Polym Sci Part A Polym Chem 27:1795–1809
85. Cho I, Kim J (1998) Polymer 40:1577–1580
86. Ishizu K, Tahara N (1996) Polymer 37:1729–1734
87. Robin JJ (2004) Adv Polym Sci 167:35–79
88. Ebdon JR, Flint NJ (1992) Polym Prepr Am Chem Soc Div Polym Chem 33:972–973

89. Cheradame H (1989) In: Goethals EJ (ed) Telechelic polymers. CRC, Boca Raton, p 147–167
90. Dix LR, Ebdon JR, Flint NJ, Hodge P (1991) Eur Polym J 27:581–588
91. Dix LR, Ebdon JR, Hodge P (1993) Polymer 34:406–411
92. Ebdon JR (1994) Macromol Symp 84:45–54
93. Ebdon JR, Flint NJ, Rimmer S (1995) Macromol Rep A 32:603–611
94. Liu Z, Ebdon J, Rimmer S (2004) React Funct Polym 58:213–224
95. Ebdon JR, Flint NJ (1996) Eur Polym J 32:289
96. Rimmer S, Ebdon JR (1996) J Polym Sci Part A Polym Chem Ed 34:3573
97. Rimmer S, Ebdon JR (1997) J Chem Res Synop 11:408
98. Asandei AD, Percec V (2001) J Polym Sci Part A Polym Chem 39:3392–3418
99. Fukuda T, Yoshikawa C, Kwak Y, Goto A, Tsujii Y (2003) ACS Symp Ser 854:24–39
100. Fukuda T, Goto A, Ohno K (2000) Macromol Rapid Commun 21:151–165
101. Fukuda T, Goto A (2000) ACS Symp Ser 768:27–38
102. Matyjaszewski K (2003) ACS Symp Ser 854:2–9
103. Matyjaszewski K (ed) (2000) Controlled/living radical polymerization. Progress in ATRP, NMP, and RAFT. Proceedings of a symposium on controlled radical polymerization held on 22–24 August 1999, in New Orleans. ACS symposium series 768. Am Chem Soc, Washington
104. Wang J-S, Matyjaszewski K (1995) J Am Chem Soc 117:5614–5615
105. Kamigaito M, Ando T, Sawamoto M (2001) Chem Rev 101:3689–3745
106. Kato M, Kamingaito M, Sawamoto M, Higashimura T (1995) Macromolecules 28:1721
107. Zhang X, Matyjaszewski K (1999) Macromolecules 32:7349–7353
108. Ando T, Kamigaito M, Sawamoto M (1998) Macromolecules 31:6708–6711
109. Nakagawa Y, Matyjaszewski K (1998) Polym J 30:138–141
110. Nakagawa Y, Gaynor SG, Matyjaszewski K (1996) Polym Prepr Am Chem Soc Div Polym Chem 37:577–578
111. Sadhu VB, Pionteck J, Voigt D, Komber H, Fischer D, Voit B (2004) Macromol Chem Phys 205:2356–2365
112. Ando T, Kamigaito M, Sawamoto M (1997) Macromolecules 30:4507–4510
113. Ando T, Kato M, Kamigaito M, Sawamoto M (1996) Macromolecules 29:1070–1072
114. Kato M, Kamigaito M, Sawamoto M, Higashimura T (1995) Macromolecules 28:1721–1723
115. Granel C, Dubois P, Jerome R, Teyssie P (1996) Macromolecules 29:8576–8582
116. Wang J-S, Gaynor SG, Matyjaszewski K (1995) Polym Prepr Am Chem Soc Div Polym Chem 36:465–466
117. Xia J, Matyjaszewski K (1997) Macromolecules 30:7697–7700
118. Haddleton DM, Crossman MC, Dana BH, Duncalf DJ, Heming AM, Kukulj D, Shooter AJ (1999) Macromolecules 32:2110–2119
119. Haddleton DM, Perrier S, Bon SAF (2000) Macromolecules 33:8246–8251
120. Persec V, Barbolu B, Bera TK, Kim HJ, Fréchet JM, Grubbs RH (1999) Preprints of IUPAC international symposium on ionic polymerization, p 37
121. Haddleton DM, Jasieczezek CB, Hannon MJ, Shooter AJ (1997) Macromolecules 30:2190–2193
122. Malz H, Komber H, Voigt D, Hopfe I, Pionteck J (1999) Macromol Chem Phys 200:642–651
123. Matyjaszewski K (1998) ACS Symp Ser 685:2–30
124. Moon B, Hoye TR, Macosko C (2001) Macromolecules 34:7941–7951
125. Fallais L, Pantoustier N, Devaux J, Zune C, Jérôme R (2000) Polymer 40:5535

126. Zhang Y, Tebby JC, Wheeler JW (1998) Eur Polym J 35:209-214
127. Paik H-J, Teodorescu M, Xia J, Matyjaszewski K (1999) Macromolecules 32:7023-7031
128. Keul H, Neumann A, Reining B, Hocker H (2000) Macromol Symp 161:63-72
129. Percec V, Kim H-J, Barboin B (1997) Macromolecules 30:6702-6705
130. Coessens V, Nakagawa Y, Matyjaszewski K (1998) Polym Bull 40:135-142
131. Coessens V, Matyjaszewski K (1999) J Macromol Sci Pure Appl Chem A 36:667-679
132. Li L, Wang C, Long Z, Fu S (2000) J Polym Sci Part A Polym Chem 38:4519-4523
133. Coessens V, Matyjaszewski K (1999) J Macromol Sci Pure Appl Chem A 36:667-679
134. Matyjaszewski K (1996) Curr Opin Solid State Mater Sci 1:769-776
135. Snijder A, Klumperman B, Van Der Linde R (2002) J Polym Sci Part A Polym Chem 40:2350-2359
136. Shim AK, Coessens V, Pintauer T, Gaynor S, Matyjaszewski K (1999) Polym Prepr Am Chem Soc Div Polym Chem 40:456-457
137. Coessens V, Matyjaszewski K (1999) Macromol Rapid Commun 20:127-134
138. Bon SAF, Morsley SR, Waterson C, Haddleton DM (2000) Macromolecules 33:5819-5824
139. Koulouri EG, Kallitsis JK, Hadziioannou G (1999) Macromolecules 32:6242-6248
140. Bon SAF, Steward AG, Haddleton DM (2000) J Polym Sci Part A Polym Chem 38:2678-2686
141. Joshi RM (1962) Macromol Chem 53:33-42
142. Coessens V, Nakagawa Y, Matyjaszewski K (1998) Polym Bull 40:135-142
143. Peters R, Mengerink JK, Langereis S, Frederix M, Linssen H, Van Hest J, Van Der Wall S (2002) J Chromatogr A 949:327-335
144. Rostovtsev VV, Green LG, Fokin VV, Sharpless KB (2002) Angew Chem Int Ed Engl 41:2596-2599
145. Kolb HC, Finn MG, Sharpless KB (2001) Angew Chem Int Ed Engl 40:2004-2021
146. Diaz DD, Punna S, Holzer P, McPherson AK, Sharpless KB, Fokin VV, Finn MG (2004) J Polym Sci Part A Polym Chem 42:4392-4403
147. Binder WH, Kluger C (2004) Macromolecules 37:9321-9330
148. Lutz J-F, Boener HG, Weichenhan K (2005) Macromol Rapid Commun 26:514-518
149. Summerlin BS, Tsarevsky NV, Louches G, Lee RY, Matyjaszewski K (2005) Macromolecules 38:7540-7545
150. Lutz J-F, Boerner HG, Weichenhan K (2005) Polym Prepr Am Chem Soc Div Polym Chem 46:486-487
151. Lutz J-F, Boerner HG, Weichenhan K (2005) Macromol Rapid Commun 26:514-518
152. Lewis WG, Magallon FG, Fokin VV, Finn MG (2004) J Am Chem Soc 126:9152-9153
153. Fukui H, Sawamoto M, Higashimura T (1993) Macromolecules 26:7315-7321
154. Tokuchi K, Ando T, Sawamato M (2000) Macromolecules 31:6708-6711
155. Percec V, Popov AV, Ramirez-Castillo E, Weichold O (2003) J Polym Sci Part A Polym Chem 41:3283-3299
156. Matyjaszewski K, Xia J (2001) Chem Rev 101:2921-2990
157. Chambard G, Klumperman B, German AL (2000) Macromolecules 33:4417-4421
158. Critchley JP, McLoughlin VCR, Thrower J, White IM (1970) Br Polym J 2:288-294
159. Kim YK, Pierce OR (1968) J Org Chem 33:442-443
160. Otazaghine B, Boutevin B (2004) Macromol Chem Phys 205:2002-2011
161. Otazaghine B, David G, Boutevin B, Robin JJ, Matyjaszewski K (2004) Macromol Chem Phys 205:154-164
162. Otazaghine B, Boyer C, Robin J-J, Boutevin B (2005) J Polym Sci Part A Polym Chem 43:2377-2394

163. Sarbu T, Lin K-Y, Spanswick J, Gil RR, Siegwart DJ, Matyjaszewski K (2004) Macromolecules 37:9694–9700
164. Yurteri S, Cianga I, Yagci Y (2003) Macromol Chem Phys 204:1771–1783
165. Chiefari J, Chong Y, Ercole F, Krstina J, Jeffery J, Le T, Mayadunne R, Meijs GF, Moad CL, Moad G, Rizzardo E, Thang SH (1998) Macromolecules 31:5559
166. Le T, Moad G, Rizzardo E, Thang SH (1998) In PCT Int Appl WO 9801478
167. Destarac M, Charmot D, Franck X, Zard S (2000) Macromol Rapid Commun 21:1035
168. Corpart P, Charmot D, Biadatti T, Zard S, Michelet D (1998) In PCT Int Appl WO 9858974
169. Delduc P, Tailhan C, Zard SZ (1988) J Chem Soc Chem Commun 308–310
170. Chiefari J, Jeffery J, Mayadunne RTA, Moad G, Rizzardo E, Thang SH (2000) ACS Symp Ser 768:297–312
171. Monteiro MJ (2005) J Polym Sci Part A Polym Chem Ed 43:3189
172. Barner-Kowollik C, Quinn JF, Morsley DR, Davis TP (2001) J Polym Sci Part A Polym Chem Ed 39:1353
173. Stenzel-Rosenbaum M, Davis TP, Chen V, Fane AG (2001) J Polym Sci Part A Polym Chem Ed 39:2777
174. Goto A, Sato K, Tsujii Y, Fukuda T, Moad G, Rizzardo E, Thang SH (2001) Macromolecules 34:402
175. Summerlin BS, Donovan MS, Mitsukami Y, Lowe AB, McCormick CL (2001) Macromolecules 34:6561
176. Monteiro MJ, de Barbeyrac J (2001) Macromolecules 34:4416
177. de Brouwer H, Tsavalas JG, Schork FJ, Monteiro MJ (2000) Macromolecules 33:9239
178. Tonge MP, McLeary JB, Vosloo JJ, Sanderson RD (2003) Macromol Symp 193:289
179. Smulders W, Monteiro MJ (2004) Macromolecules 37:4474
180. Prescott SW, Ballard MJ, Rizzardo E, Gilbert RG (2002) Macromolecules 35:5417
181. Postma A, Davis TP, Moad G, O'Shea MS (2005) Macromolecules 38:5371–5374
182. Theis A, Feldermann A, Charton N, Stenzel MH, Davis TP, Barner-Kowollik C (2005) Macromolecules 38:2595–2605
183. Llauro M-F, Loiseau J, Boisson F, Delolme F, Ladaviere C, Claverie J (2004) J Polym Sci Part A Polym Chem 42:5439–5462
184. Baussard J-F, Habib-Jiwan J-L, Laschewsky A, Mertoglu M, Storsberg J (2004) Polymer 45:3615–3626
185. McCormick CL, Donovan MS, Lowe AB, Sumerlin BS, Thomas DB (2003) US Patent Appl 2 003 195 310
186. McCormick CL, Donovan MS, Lowe AB, Sumerlin BS, Thomas DB (2003) PCT Int Appl 2003066685
187. Liu J, Hong C-Y, Pan C-Y (2004) Polymer 45:4413–4421
188. Liu RCW, Segui F, Viitala T, Winnik FM (2004) PMSE Prepr 90:105–106
189. Lai JT, Filla D, Shea R (2002) Macromolecules 35:6754–6756
190. Lima V, Brokken-Zijp J, Klumperman B, van Benthem-van Duuren G, van der Linde R (2003) Polym Prepr Am Chem Soc Div Polym Chem 44:812–813
191. Lima VGR, Brokken J, Klumperman B, van Benthem-van Duuren G, van der Linde R (2003) Abstracts of papers, 225th ACS national meeting, New Orleans, LA, United States, March 23–27, 2003, POLY-047
192. Lima V, Jiang X, Brokken-Zijp J, Schoenmakers PJ, Klumperman B, Van Der Linde R (2005) J Polym Sci Part A Polym Chem 43:959–973
193. Entelis SG, Evreinov VV, Gorshkov AV (1986) Adv Polym Sci 76:129
194. Cools PJCH, van Herk AM, German AL, Staal W (1994) J Liq Chromatogr 17:3133
195. Macko T, Hunkeler D (2003) Adv Polym Sci 163:61

196. Baek KY, Kamigaito M, Sawamoto M (2002) J Polym Sci Part A Polym Chem Ed 40:1937
197. Jiang X, Schoenmakers Peter J, Lou X, Lima V, van Dongen Joost LJ, Brokken-Zijp J (2004) J Chromatogr A 1055:123–133
198. Jiang S, Viehe HG, Oger N, Charmot D (1995) Macromol Chem Phys 196:2349
199. Convertine AJ, Lokitz BS, Lowe AB, Scales CW, Myrick LJ, McCormick CL (2005) Macromol Rapid Commun 26:791–795
200. Lewandowski KM, Fansler DD, Wendland MS, Heilmann SM, Gaddam BN (2004) US Patent 6 762 257
201. Lewandowski KM, Fansler DD, Wendland MS, Gaddam BN, Heilmann SM (2004) Patent US 6 753 391
202. Perrier S, Takolpuckdee P, Mars CA (2005) Macromolecules 38:2033–2036
203. Solomon DH, Rizzardo E, Cacioli P (1985) Eur Patent 135 280
204. Georges MK, Veregin RPN, Kazmaier PM, Hamer GK (1993) Macromolecules 26:2987
205. Georges MK, Veregin RPN, Kazmaier PM, Hamer GK (1994) Macromolecules 27:7228
206. Matyjaszewski K, Gaynor S, Greszta D, Mardare D, Shigemoto T (1995) J Phys Org Chem 8:306
207. Hawker CJ (1994) J Am Chem Soc 116:11185
208. Goto A, Fukuda T (1997) Macromolecules 30:4272–4277
209. Fukuda T, Tsujii Y, Miyamoto T (1997) Polym Prepr Am Chem Soc Div Polym Chem 38:723–724
210. Fukuda T, Terauchi T, Goto A, Ohno K, Tsujii Y, Miyamoto T, Kobatake S, Yamada B (1996) Macromolecules 29:6393–6398
211. Benoit D, Harth E, Fox P, Waymouth RM, Hawker CJ (2000) Macromolecules 33:363–370
212. Benoit D, Chaplinski V, Braslau R, Hawker CJ (1999) J Am Chem Soc 121:3904–3920
213. Veregin RPN, Georges MK, Kazmaier PM, Hamer GK (1993) Macromolecules 26:5316
214. Rizzardo E (1987) Chem Aust 54:32
215. Georges MK, Veregin RPN, Kazmaier PM, Hamer GK (1993) Macromolecules 26:2987–2988
216. Pradel J-L, Ameduri B, Boutevin B (1999) Macromol Chem Phys 200:2304–2308
217. Pradel J-L, Ameduri B, Boutevin B, Lacroix-Desmazes P (1999) Polym Prepr Am Chem Soc Div Polym Chem 40:382–383
218. Pradel JL, Boutevin B, Ameduri B (2000) J Polym Sci Part A Polym Chem 38:3293–3302
219. Boutevin B, Cerf M, Pradel J-L (1997) PCT Int Appl 9746593
220. Hawker CJ, Hedrick JL (1995) Macromolecules 28:2993–2995
221. Li IQ, Howell BA, Koster RA, Priddy DB (1996) Macromolecules 29:8554–8555
222. Hammouch SO, Catala JM (1996) Macromol Rapid Commun 17:149
223. Harth E, Hawker CJ, Fan W, Waymouth RM (2001) Macromolecules 34:3856–3862
224. Lohmeijer BGG, Schubert US (2004) J Polym Sci Part A Polym Chem 42:4016–4027
225. Lohmeijer BGG, Schubert US (2002) Angew Chem Int Ed Engl 41:3825–3829
226. Tatemoto M, Suzuki T, Tomoda M, Furukawa Y, Ueta Y (1978) Ger Offen 2815187
227. Tatemoto M, Nakagawa T (1986) Jpn Tokkyo Koho 61049327
228. Tatemoto M, Yutani Y, Fujiwara K (1988) Eur Patent Appl 272 698
229. Tatemoto M (1992) Kobunshi Ronbunshu 49:765–783
230. Greszta D, Mardare D, Matyjaszewski K (1994) Macromolecules 27(3):638–644

231. Hung MH (1993) US Patent 5 231 154
232. Arcella V, Brinati G, Apostolo M (1997) Chim Ind 79:345–351
233. Percec V, Popov AV, Ramirez-Castillo E, Coelho JFJ, Hinojosa-Falcon LA (2004) J Polym Sci Part A Polym Chem 42:6267–6282
234. Percec V, Popov AV (2005) J Polym Sci Part A Polym Chem 43:1255–1260
235. Percec V, Popov AV, Ramirez-Castillo E, Coelho JFJ (2005) J Polym Sci Part A Polym Chem 43:773–778
236. Percec V, Popov AV, Ramirez-Castillo E, Monteiro M, Barboiu B (2002) J Am Chem Soc 124:4940
237. Percec V, Popov AV, Ramirez-Castillo E, Weichold O (2003) J Polym Sci Part A Polym Chem Ed 41:3283
238. Ameduri B, Boutevin B (2004) Well-architectured fluoropolymers: synthesis, properties and applications. Elsevier, Kidlington
239. Feiring AE (1994) J Macromol Sci Pure Appl Chem A 31:1657–1673
240. McLoughlin VCR, Thrower J (1969) Tetrahedron 25:5921–5940
241. Baum K (1992) Synth Fluor Chem 381–393
242. McLoughlin VC, Thrower J (1970) Br Patent 1 208 451
243. McLoughlin VCR, Thrower J (1968) US Patent 3 408 411
244. Ameduri B, Boutevin B, Kostov G (2001) Prog Polym Sci 26:105–187
245. Boulahia D, Manseri A, Ameduri B, Boutevin B, Caporiccio G (1999) J Fluor Chem 94:175–182
246. Manseri A, Ameduri B, Boutevin B, Kotora M, Hajek M, Caporiccio G (1995) J Fluor Chem 73:151–158
247. Lahiouhel D, Ameduri B, Boutevin B (2001) J Fluor Chem 107:81–88
248. Ameduri B, Boutevin B, Guida-Pietrasanta F, Manseri A, Ratsimihety A, Caporiccio G (1996) J Polym Sci Part A Polym Chem 34:3077–3090
249. Ameduri B, Boutevin B, Caporiccio G, Guida-Pietrasanta F, Manseri A, Ratsimihety A (1999) Fluoropolymers 1:67–80
250. Ameduri B, Colomines G, Rousseau A, Boutevin B, Andre S, Andrieu X (2003) In: Fluorine in coatings V, conference Papers, 5th, Orlando, FL, United States, January 21–22, 2003. Paper19/M, Paper19/11–Paper19/22
251. Tsukahara Y, Tsutsumi K, Yamashita Y, Shimada S (1989) Macromolecules 22:2869–2871
252. Teodorescu M (2001) Eur Polym J 37:1417–1422
253. Boyer C, Loubat C, Robin JJ, Boutevin B (2004) J Polym Sci Part A Polym Chem 42:5146–5160
254. Chen GF, Jones FN (1991) Macromolecules 24:2151–2155
255. Boyer C, Boutevin G, Robin JJ, Boutevin B (2004) Macromol Chem Phys 205:645–655
256. Oishi T, Lee Y-K, Nakagawa A, Onimura K, Tsutsumi H (2002) J Polym Sci Part A Polym Chem 40:1726–1741
257. Chen M-Q, Kishida A, Akashi M (1996) J Polym Sci Part A Polym Chem 34:2213–2220
258. Akashi M (1996) Jpn Kokai Tokkyo Koho 08183760
259. Serizawa T, Chen M-Q, Akashi M (1998) Langmuir 14:1278–1280
260. Serizawa T, Chen M-Q, Akashi M (1998) J Polym Sci Part A Polym Chem 36:2581–2587
261. Chen M-Q, Serizawa T, Akashi M (1999) Polym Adv Technol 10:120–126
262. Chen C-W, Serizawa T, Akashi M (1999) Langmuir 15:7998–8006
263. Chen M, Chen Y, Liu X, Yang C, Akashi M (2002) Gaofenzi Xuebao 447–451

264. Serizawa T, Matsukuma D, Nanameki K, Uemura M, Kurusu F, Akashi M (2004) Macromolecules 37:6531–6536
265. Serizawa T, Uemura M, Kaneko T, Akashi M (2002) J Polym Sci Part A Polym Chem 40:3542–3547
266. Suzuki K, Yumura T, Mizuguchi M, Tanaka Y, Chen C-W, Akashi M (2000) J Appl Polym Sci 77:2678–2684
267. Seto F, Fukuyama K, Muraoka Y, Kishida A, Akashi M (1998) J Appl Polym Sci 68:1773–1779
268. Wood CD, Cooper AI (2003) Macromolecules 36:7534–7542
269. Ohnaga T, Sato T (1996) Polymer 37:3729–3735
270. Fukutomi T, Ishizu K, Shiraki K (1987) J Polym Sci Part C Polym Lett 25:175–180
271. Chen W, Kobayashi S, Inoue T, Ohnaga T, Ougizawa T (1994) Polymer 35:4015–4021
272. Shigehisa T, Akasu H, Onaga T, Sato T (1992) Jpn Kokai Tokkyo Koho 04305233
273. Onaga T, Fukushima T, Otsuka K, Sato T (1992) Jpn Kokai Tokkyo Koho 04296347
274. Sato T, Onaga T, Ikeda K (1992) Jpn Kokai Tokkyo Koho 04139203
275. Sato T, Ohnaga T, Ikeda K (1991) Eur Patent Appl 421 296
276. Collins S, Rimmer S (2002) Polym Prepr Am Chem Soc Div Polym Chem 43:1085–1086
277. Carter S, Kavros A, Rimmer S (2001) React Funct Polym 48:97–105
278. Collins S, Rimmer S (2004) Rapid Commun Mass Spectrom 18:3075–3078
279. Nair PR, Nair CPR, Francis DJ (1997) Eur Polym J 33:89–95
280. Chujo Y, Tatsuda T, Yamashita Y (1982) Polym Bull 8:239–244
281. Jayakumar R, Nanjundan S, Prabaharan M (2005) J Macromol Sci Polym Rev C 45:231–261
282. Takeichi T, Guo Y, Agag T (2000) J Polym Sci Part A Polym Chem 38:4165–4176
283. Yamashita Y, Chujo Y, Kobayashi H, Kawakami Y (1981) Polym Bull 5:361–366
284. Chujo Y, Kobayashi H, Yamashita Y (1984) Polym Commun 25:278–280
285. Chujo Y, Shishino T, Tsukahara Y, Yamashita Y (1985) Polym J 17:133–141
286. Chujo Y, Kobayashi H, Yamashita Y (1988) Polym J 20:407–411
287. Chujo Y, Hiraiwa A, Kobayashi H, Yamashita Y (1988) J Polym Sci Part A Polym Chem 26:2991–2996
288. Chujo Y, Kohno K, Usami N, Yamashita Y (1989) J Polym Sci Part A Polym Chem 27:1883–1890
289. Okamoto M (2001) J Appl Polym Sci 80:2670–2675
290. Kim D-K, Lee S-B, Doh K-S, Nam Y-W (1999) J Appl Polym Sci 74:2029–2038
291. Kim D-K, Lee S-B, Doh K-S, Nam Y-W (1999) J Appl Polym Sci 74:1917–1926
292. Meijs GF, Rizzardo E, Thang SH (1990) Polym Bull 24:501
293. Yagci Y, Reetz I (1999) React Funct Polym 42:255–264
294. Yamada B, Tagashira S, Aoki S (1994) J Polym Sci Part A Polym Chem 32:2745–2754
295. Yamada B, Kobetake S (1994) Prog Polym Sci 19:1089
296. Burczyk AF, O'Driscoll KF, Rempel GL (1984) J Polym Sci Polym Chem Ed 22:3255–3262
297. Cacioli BH, O' Driscoll KF, Caslett RC, Rizzardo E, Solomon DH (1986) J Macromol Sci Chem A 23:839
298. Meijs GF, Rizzardo E (1990) J Macromol Sci Rev Macromol Chem Phys C 30:305–377
299. Bamford C, Jenkins A, White EFT (1959) J Polym Sci 34:271
300. Tanaka H, Kawa H, Sato T, Ota T (1989) J Polym Sci 27:1741
301. Krstina J, Moad G, Rizzardo E, Berge CT, Fryd M (1995) Macromolecules 28:5381
302. Wang W, Stenson PA, Marin-Becerra A, McMaster J, Schroeder M, Irvine DJ, Freeman D, Howdle SM (2004) Macromolecules 37:6667–6669

303. Wang W, Stenson PA, Irvine DJ, Howdle SM (2004) PMSE Prepr 91:1051–1052
304. Pierik SCJ, van Herk AM (2004) J Appl Polym Sci 91:1375–1388
305. Kurmaz SV, Bubnova ML, Perepelitsina EO, Estrina GA (2005) Vysokomolekulyarnye Soedineniya, Seriya A 47:414–429
306. Gridnev AA, Ittel SD (1996) Macromolecules 29:5864–5874
307. Gridnev AA, Ittel SD (1999) Book of abstracts, 218th ACS national meeting, New Orleans, August 22–26. POLY-452
308. Davis TP, Haddleton DM, Richards SN (1994) J Macromol Sci Rev Macromol Chem Phys C34:243
309. Davis TP, Kukulj D, Haddleton DM (1995) Trends Polym Sci 3:365
310. Darmon M, Berge C, Antonelli J (1993) US Patent 5 264 530
311. Suddaby KG, Sanayei R, O'Driscoll K (1991) J Appl Polym Sci 43:1565
312. Li Y, Wayland BB (2003) Chem Commun 1594–1595
313. Li Y, Lu Z, Wayland BB (2003) Polym Prepr Am Chem Soci Div Polym Chem 44:780
314. Chiu TYJ, Heuts JPA, Davis TP, Stenzel MH, Barner-Kowollik C (2004) Macromol Chem Phys 205:752–761
315. Chiefari J, Jeffery J, Moad G, Rizzardo E, Thang SH (1999) Polym Prepr Am Chem Soc Div Polym Chem 40:344–345
316. Chiefari J, Jeffery J, Mayadunne RTA, Moad G, Rizzardo E, Thang SH (1999) Macromolecules 32:7700–7702
317. Chiefari J, Jeffery J, Moad G, Rizzardo E, Thang SH (1999) Book of abstracts, 218th ACS national meeting, New Orleans, August 22–26. POLY-317
318. Matyjaszewski K, Beers KL, Kern A, Gaynor SG (1998) J Polym Sci Part A Polym Chem 36:823–830
319. Zeng F, Shen Y, Zhu S, Pelton R (2000) Macromolecules 33:1628–1635
320. Neugebauer D, Zhang Y, Pakula T, Matyjaszewski K (2003) Polymer 44:6863–6871
321. Schoen F, Hartenstein M, Mueller AHE (2001) Macromolecules 34:5394–5397
322. Cheng G, Simon PFW, Hartenstein M, Muller AHE (2000) Macromol Rapid Commun 21:846–852
323. Hua F, Kita R, Wegner G, Meyer W (2005) Chem Phys Chem 6:336–343
324. Muehlebach A (2004) PMSE Prepr 90:180
325. Schulze U, Fonagy T, Komber H, Pompe G, Pionteck J, Ivan B (2003) Macromolecules 36:4719–4726
326. Kennedy JP, Ivan B (1992) Designed polymers by carbocationic macromolecular engineering. Theory and practice. Hanser publisher, Munich
327. Isasi JR, Mandelkern L, Galante MJ, Alamo RG (1999) J Polym Sci Part B Polym Phys 37:323–334
328. Ivan B, Fonagy T (1999) Polym Prepr Am Chem Soc Div Polym Chem 40:356–357
329. Couvreur L, Sharma B, Du Prez F (2005) Polym Prepr Am Chem Soc Div Polym Chem 46:462–463
330. Couvreur L, Sharma B, Du Prez FE (2005) Abstracts of papers, 230th ACS national meeting, Washington, DC, United States, August 28-September 1, 2005. POLY-450
331. Norman J, Moratti SC, Slark AT, Irvine DJ, Jackson AT (2002) Macromolecules 35:8954–8961
332. Mecerreyes D, Dahan D, Lecomte P, Dubois P, Demonceau A, Noels AF, Jerome R (1999) J Polym Sci Part A Polym Chem 37:2447–2455
333. Mecerreyes D, Atthoff B, Boduch KA, Trollsaas M, Hedrick JL (1999) Macromolecules 32:5175–5182
334. Mecerreyes D, Pomposo JA, Bengoetxea M, Grande H (2000) Macromolecules 33:5846–5849

335. Malz H, Pionteck J, Potschke P, Komber H, Voigt D, Luston J, Bohme F (2001) Macromol Chem Phys 202:2148–2154
336. Cianga I, Yagci Y (2004) Prog Polym Sci 29:387–399
337. Wagner M, Nuyken O (2004) J Macromol Sci Pure Appl Chem A 41:637–647
338. Karavia V, Deimede V, Kallitsis JK (2004) J Macromol Sci Pure Appl Chem A 41:115–131
339. Deimede V, Kallitsis JK (2002) Chem Eur J 8:467–473
340. Deimede V, Kallitsis Joannis K (2002) Chemistry 8:467–473
341. Kuckling D, Wohlrab S (2001) Polymer 43:1533–1536
342. Ding J, Chuy C, Holdcroft S (2002) Macromolecules 35:1348–1355
343. Hawker CJ, Mecerreyes D, Elce E, Dao J, Hedrick JL, Barakat I, Dubois P, Jerome R, Volksen I (1997) Macromol Chem Phys 198:155–166
344. Hawker CJ, Carter KR, Hedrick JL, Volksen W (1995) Polym Prepr Am Chem Soc Div Polym Chem 36:110–111
345. Hong H, Chung TC (2004) Macromolecules 37:6260
346. Leibler L (2005) Prog Polym Sci 30:898–914
347. Schwindeman JA, Letchford RJ, Granger EJ, Quirk RP (2002) PCT Int Appl 2002060958
348. Coessens V, Pyun J, Miller PJ, Gaynor S, Matyjaszewski K (2000) Macromol Rapid Commun 21:103
349. Bamford C, Jenkins A, Johnston R (1959) Trans Faraday Soc 55:1451
350. Olaj O (1971) Makromol Chem 15:249
351. Deb P, Meyerhoff G (1974) Eur Polym J 10:709
352. Mahabadi H, Meyerhoff G (1978) Eur Polym J 15:607
353. Ito K (1969) J Polym Sci Part A Polym Chem Ed 7:2995
354. Vertommen LLT, Meijer J (1991) PCT Int Appl WO 9107440
355. Meijs GF, Rizzardo E, Thang SH (1992) Polym Prepr 33:893
356. Colombani D, Chaumont P (1995) Polymer 36:129–136
357. Meijs GF, Rizzardo E (1991) Polym Int 26:239
358. Bailey WJ, Endo T, Gapud B, Lin YN (1984) J Macromol Sci Chem A 21:979
359. Zink M-O, Colombani D, Chaumont P (1997) Eur Polym J 33:1433
360. Sato E, Zetterlund PB, Yamada B (2004) J Polym Sci Part A Polym Chem 42:6021–6030
361. Otsu T, Matsumoto A, Tazaki T (1987) Polym Bull 17:323–330
362. Otsu T, Matsumoto A, Tazaki T (1986) Mem Fac Eng Osaka City Univ 27:137–142
363. Borsig E, Lazar M, Capla M (1967) Makromol Chem 105:212–222
364. Borsig E, Lazar M, Capla M, Florian S (1969) Angew Makromol Chem 9:89–95
365. Bledzki A, Braun D (1981) Makromol Chem 182:1047–1056
366. Bledzki A, Balard H, Braun D (1981) Makromol Chem 182:3195–3206
367. Balard H, Bledzki A, Braun D (1981) Makromol Chem 182:1063–1071
368. Bledzki A, Balard H, Braun D (1981) Makromol Chem 182:1057–1062
369. Roussel J, Boutevin B (2001) J Fluor Chem 108:37–45
370. Odinokov VN, Kukovinets OS (1978) Otkrytiya Izobret Prom Obraztsy Tovarnye Znaki 88:101
371. Tikhomirov BI, Baraban OP (1969) Vysokomol Soedin 70:759
372. Dubois DA (2000) WO Patent 2 000 032 645
373. Godt HC (1967) Fr Patent 1 497 289
374. Godt HC (1969) US Patent 3 429 936
375. Weider R, Scholl T, Kohler B (1996) US Patent 5 484 857

376. D'Agosto F, Hughes R, Charreyre M-T, Pichot C, Gilbert RG (2003) Macromolecules 36:621–629
377. Kobetake S, Hardwood H, Quirk RP, Priddy DB (1998) J Polym Sci Part A Polym Chem Ed 36:2555
378. Li IQ, Knauss DM, Priddy DB, Howell BA (2003) Polym Int 52:805–812
379. Rodlert M, Harth E, Rees I, Hawker CJ (2000) J Polym Sci Part A Polym Chem 38:4749–4763
380. Rolland JP, DeSimone JM (2003) PMSE Prepr 88:606–607
381. Desimone JM, Rolland J (2003) Abstracts of papers, 225th ACS national meeting, New Orleans, LA, United States, March 23–27, 2003. PMSE-350
382. Se K, Aoyama K (2004) Polymer 45:79–85
383. Se K, Aoyama K, Aoyama J, Donkai M (2003) Macromolecules 36:5878–5881
384. Kim J-H, Kim J-G, Kim D, Kim YH (2005) J Appl Polym Sci 96:56–61
385. Gallot B, Douy A (1987) Mol Cryst Liq Cryst 153:367–373
386. Nguyen S, Marchessault RH (2005) Macromolecules 38:290–296
387. Nguyen S, Marchessault RH (2004) Macromol Biosci 4:262–268
388. Eguiburu JL, Fernandez-Berridi MJ, San Roman J (2000) Polymer 41:6439–6445
389. Eguiburu JL, Fernandez-Berridi MJ, San Roman J (1996) Polymer 37:3615–3622
390. Lutz J-F, Jahed N, Matyjaszewski K (2004) J Polym Sci Part A Polym Chem 42:1939–1952
391. Shinoda H, Matyjaszewski K (2001) Macromolecules 34:6243–6248
392. Yan-Ming G, Ting W, Yin-Fang Z, Cai-Yuan P (2001) Polymer 42:6385–6391
393. Sierra-Vargas J, Masson P, Beinert G, Rempp P, Franta E (1982) Polym Bull 7:277–282
394. Larraz E, Elvira C, Gallardo A, San Roman J (2005) Polymer 46:2040–2046
395. Da Cunha C, Deffieux A, Fontanille M (1992) J Appl Polym Sci 44:1205–1212
396. Da Cunha C, Deffieux A, Fontanille M (1993) J Appl Polym Sci 48:819–831
397. Kobayashi K, Kamiya S, Enomoto N (1996) Macromolecules 29:8670–8676
398. Wataoka I, Urakawa H, Kobayashi K, Akaike T, Schmidt M, Kajiwara K (1999) Macromolecules 32:1816–1821
399. Ishizu K, Tsubaki K, Uchida S (2002) Macromolecules 35:10193–10197
400. Tsubaki K, Ishizu K (2001) Polymer 42:8387–8393
401. Chuy C, Ding J, Swanson E, Holdcroft S, Horsfall J, Lovell KV (2003) J Electrochem Soc 150:E271–E279
402. Shen Y, Zhu S, Zeng F, Pelton R (2000) Macromolecules 33:5399–5404
403. Berlinova IV, Panayotov IM (1987) Makromol Chem 188:2141–2150
404. Lahitte J-F, Pelascini F, Peruch F, Meneghetti SP, Lutz PJ (2002) C R Chim 5:225–234
405. Feast WJ, Gibson VC, Johnson AF, Khosravi E, Mohsin MA (1997) J Mol Catal A 115:37–42
406. Nair CPR (1992) Eur Polym J 28:1527–1532
407. Meijs GF, Rizzardo E, Thang SH (1988) Macromolecules 21:3122
408. Sato K, Zetterlund PB, Yamada B (2004) J Polym Sci Part A Polym Chem Ed 42:6021

Editor: Timothy E. Long

Magnetic Field-Responsive Smart Polymer Composites

Genovéva Filipcsei · Ildikó Csetneki · András Szilágyi · Miklós Zrínyi (✉)

Department of Physical Chemistry, Budapest University of Technology and Economics,
MTA-BME Laboratory of Soft Matters, 1521 Budapest, Hungary
zrinyi@mail.bme.hu

1	Introduction	139
2	Towards Magnetic Polymer Composites	141
3	Preparation of Magnetic Polymer Gels and Elastomers	143
3.1	Preparation of Magnetic Polymer Gels with Uniform Filler Distribution	143
3.1.1	Magnetite-Loaded Poly(vinyl alcohol) Gels	143
3.1.2	Magnetic Poly(dimethylsiloxane) Elastomers with Uniform Filler Distribution	144
3.2	Preparation of Anisotropic Polymer Composites	145
3.3	Preparation of Magnetic Gel Beads	147
3.3.1	Preparation of Magnetic Polystyrene Latex	147
3.3.2	Preparation of Millimeter-Sized PNIPA and Magnetic mPNIPA Thermosensitive Gel Beads	148
3.3.3	Preparation of Magnetic Polystyrene Latex Covered by PNIPA Gel Shell	149
4	Magnetic Properties of Ferrogels	150
5	Elastic Properties of Magnetic Composites Studied by Unidirectional Compression	153
5.1	Anisotropic Mechanical Behavior	155
5.2	Anisotropic mPVA Gels and mPDMS Networks at Large Deformation	156
6	Effect of Uniform Magnetic Field on the Elastic Modulus	159
6.1	Mechanical Measurements Under Static Magnetic Field	160
6.2	Compressive Force is Perpendicular to the Direction of Particle Chains	161
6.3	Compressive Force is Parallel to the Direction of Particle Chains	162
6.4	Magnetic Elastomers in Uniform Field. A Theoretical Approach	163
7	Non-uniform Magnetic Field-Induced Deformation	166
7.1	Interpretation of Non-continuous Shape of Transition	171
7.2	Non-homogeneous Deformation	173
8	Effect of the Magnetic Particles on the Swelling Behavior of PNIPA and PDMS Gels	178
8.1	Temperature Sensitivity of PNIPA and mPNIPA Gel Beads	178
8.2	Effect of the Magnetic Particles on Swelling Kinetics of mPDMS Elastomers	181
8.3	Swelling Kinetics of Unloaded and Magnetite-Loaded PDMS Networks	183
9	Summary of Main Results	185
	References	187

Abstract The combination of polymers with nano- or microsized solid materials displays novel and often enhanced properties compared to the traditional materials. They can open up possibilities for new technological applications. Materials whose physical properties can be varied by application of magnetic fields belong to a specific class of smart materials. The broad family of magnetic field-controllable soft materials includes ferrofluids, magneto-rheological fluids, magnetic gels, and magnetic elastomers. The magnetic gels and elastomers (magnetoelasts) represent a new type of composite and consist of small magnetic particles, usually in the nanometer to micron range, dispersed in a highly elastic polymeric matrix. The magnetic particles can be incorporated into the elastic body either randomly or in ordered structure. If a uniform magnetic field is applied to the reactive mixture during the cross-linking process, particle chains form and become locked into the elastomer. The resulting composites exhibit anisotropic properties.

Combination of magnetic and elastic properties leads to a number of striking phenomena that are exhibited in response to impressed magnetic fields. The magnetic particles couple the shape and the elastic modulus with the external magnetic field. Giant deformational effects, high elasticity, anisotropic elastic and swelling properties, and quick response to magnetic fields open new opportunities for using such materials for various applications. Since the magnetic fields are convenient stimuli from the point of signal control, the magnetoelasts are promising smart materials in engineering due to their real-time controllable elastic properties.

More recently, increasing interest has been devoted to exploration of multiresponsive magnetic polymers, which exhibit sensitivity to several external stimuli. Micro- and nanospheres that combine both magnetic, temperature, and pH sensitivity were also elaborated and studied. These new results provide novel possibilities for preparation of more complex magnetic field-responsive materials like membranes with on/off switching control.

In this article, we review recent advances in mechanical and swelling behavior of magnetic field-responsive soft materials, including flexible polymer networks and gels.

Keywords Anisotropic elastomers · Ferrogels · Magnetic composites · Magnetic latexes · Stress–strain dependence · Temporary reinforcement · Vibration and shock absorber

Abbreviations

PVA	Poly(vinyl alcohol)
mPVA	Magnetite-loaded PVA
GDA	Glutaraldehyde
VA	Vinyl alcohol
PDMS	Poly(dimethylsiloxane)
mPDMS	Magnetic poly(dimethylsiloxane)
mPS	Magnetic polystyrene
SLS	Sodium lauryl sulphate
SA	Stearyl alcohol
AIBN	N,N'-Azobis(isobutyronitrile)
PNIPA	Poly(N-isopropylacrylamide)
mPNIPA	Magnetic poly(N-isopropylacrylamide)
BA	Methylenbisacrylamide
APS	Ammonium persulphate
TEMED	N,N,N',N'-Tetramethylenediamine
SEM	Scanning electron microscope

TEM	Transmission electron microscope
M	Magnetization
H	Magnetic field intensity/external field
M_s	Saturation magnetization
T_b	Blocking temperature
τ_N, τ_B	Relaxation times
ϕ_m	Volume fraction of the magnetic particles in the whole gel
μ_0	Magnetic permeability of the vacuum
m	Giant magnetic moment of nanosized magnetic particles
k_B	Boltzmann constant
T	Temperature
W_{el}	Strain energy density
$\lambda_x, \lambda_y, \lambda_z$	Deformation ratio
h_0	Initial length
h	Deformed length
G	Elastic modulus
σ_n	Nominal stress
G_0	Modulus of gel without colloidal filler particles
C_1, C_2	Mooney–Rivlin parameters
f_c	Compressive force
G_a	Apparent elastic modulus
B	Magnetic induction
$W_M(H_{eff})$	Magnetic energy contribution
$W_{el}(\lambda_z)$	Elastic energy contribution
H_{eff}	Effective magnetic field strength
W_M	Magnetic energy density
G_M^E	Magnetically induced excess modulus
f_m	Force density
χ	Initial susceptibility of the magnetoelast
G	Shear modulus of the gel
H	Magnetic field strength
q_r	Relative swelling degree
r_{10}	Radius of the gel bead at 10 °C
r_T	Radius of the gel bead at T arbitrary temperature
LCST	Lower critical solution temperature
T_C	Collapse transition temperature
AC	Anisometric quotient

1
Introduction

It is one of today's challenging tasks to manufacture new multifunctional smart materials that posses intelligence at the material level. We refer to material intelligence in terms of three main functions: *sensing* changes in environmental conditions, *processing* the sensed information, and finally *making judgment (actuating)* by moving away from or to the stimulus [1, 2]. Polymer gels are unique intelligent materials in the sense that no other class of materials can be made to respond to as many different stimuli. Volume phase transition in response to infinitesimal change of external stimuli has been observed in various gels [2–7]. The stimuli that have been demonstrated to induce abrupt changes in physical properties are diverse, and include temperature, pH, solvent or ionic composition, electric field, light intensity, as well as introduction of specific ions [2–7].

In recent years, these gels have become of major interest as novel intelligent or smart materials. Many kinds of such gels have been developed and studied for application to several biomedical and industrial fields, e.g., controlled drug delivery systems, muscle-like soft linear actuators, biomimetic energy transducing devices, and separation techniques [2–5]. Attempts at developing stimuli-responsive gels for technological purposes are complicated by the fact that structural changes, like changes in the degree of swelling, are kinetically restricted by the collective diffusion of chains and the friction between the polymer network and the swelling agent. This disadvantage often hinders the effort of designing optimal gels for different applications. In order to accelerate the response of an adaptive gel to stimuli, the use of magnetic field-sensitive gels (ferrogels) has been developed [27–42]. Combination of magnetic and elastic properties leads to a number of striking phenomena in response to impressed magnetic fields. Magnetic field-responsive materials are specific subsets of smart materials that can adaptively change their physical properties due to external magnetic field.

Magnetic materials have found a wide range of applications in science and technology. Composite materials consisting of rather rigid polymeric matrices filled with magnetic particles have been known for a long time. These materials are successfully used as permanent magnets, magnetic cores, and as connecting and fixing elements in many areas. These traditional magnetic elastomers have low flexibility and practically do not change their size, shape, or elastic properties in the presence of an external magnetic field.

Molecular or polymer magnets are systems where permanent magnetization and magnetic hysteresis can be achieved as a purely one-molecule phenomenon. Molecular magnets belong to a field that is still at an early stage of development. Their magnetic properties appear at extremely low temperature and the magnetic response is rather weak [8–10]. Another possibility for the development of magnetic polymers is to apply flexible polymer composites containing magnetic particles. Magnetic elastomers (magnetoelasts) [11–26] as well as magnetic gels (ferrogels) [27–49] have been developed recently. A relatively limited number of comprehensive studies have been devoted to understanding the coupling phenomena between magnetic and elastic properties, despite the fact that several attempts have been made to apply magnetic elastomers as soft actuators, micromanipulators, artificial muscles, tuneable or adjustable mounts, and as suspension devices [47–60].

The variety of magnetic polymers is summarized in Fig. 1.

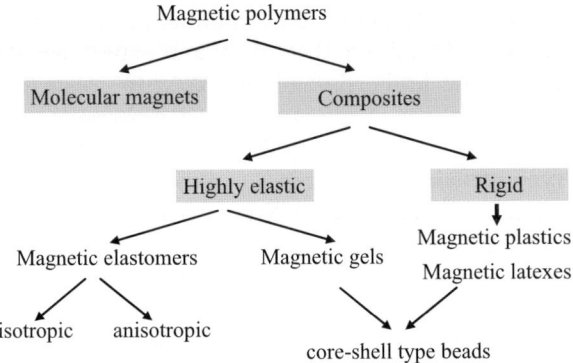

Fig. 1 Magnetic polymers

The main purpose of the present review is to report on recent advances in the development of magnetic field-responsive polymer composites. The review article is organized as follows. In the first part the diversity of magnetic polymers is outlined. The next section summarizes the preparation of magnetic polymer gels and elastomers, including monolith samples and microspheres. Preparation of isotropic composites containing randomly distributed magnetic particles, as well as highly anisotropic samples characterized by ordered structure of filler particles, is then described. The magnetic properties of these materials are the next topic of discussion. This is followed by the discussion of stress–strain behavior. Section 6 deals with the influence of a uniform magnetic field on the elastic modulus. In the next two sections, the magnetoelastic coupling is discussed. A final section is concerned with the swelling behavior of magnetite-loaded polymer gels.

2
Towards Magnetic Polymer Composites

Many useful engineering materials, as well as living organisms have a heterogeneous composition. The hybridization of organic and inorganic matter on the colloidal scale provides new and sometimes surprising properties. Fillers are usually solid additives that are incorporated into the polymer to modify the physical properties. Fillers can be divided into three categories: First are those that reinforce the polymer and improve its mechanical performance. Second are those that are used to take up space, and thus reduce material cost. The third, less common category is when filler particles are incorporated into the material to improve its responsive properties.

The new generation of magnetic elastomers and gels represent a new type of composites, consisting of small (mainly nano- and micron-sized) magnetic particles dispersed in a high elastic polymeric matrix [11–51]. These materials are relatively new and exhibit a great number of fascinating phenomena, which are the subject of intensive theoretical and experimental research. The peculiar magnetoelastic properties may be used to create a wide range of motions and to control the shape change and movement. An understanding of magneto-elastic coupling in polymers will hasten the engineers to develop new type of switches, sensors, micromachines, biomimetic energy-transducing devices, and controlled delivery systems.

Magnetically active soft materials are polymer-based elastomeric materials that react to an external magnetic field and undergo deformation or experience mechanical stress. They are often cited by various synonyms as magnetostrictive polymers, magnetorheological polymers, or magnetoelasts. Magnetic polymer gels belong to a subclass of magnetic elastomers where the flexible cross-linked polymers contain magnetizable particles as well as a significant amount of swelling liquid.

Development of magnetically active polymer systems is strongly related to the development of magnetic nanoparticles and magnetic fluids [52–58]. Magnetic nanoparticles are of considerable interest because of their potential use in high-density memory devices, spintronics, and application in diagnostic medicine.

Magnetic liquids (ferrofluids) are colloidal system of single domain magnetic nanoparticles that are dispersed either in aqueous or organic liquids [52–58]. The particles are typically in the size range 5–15 nm and are held in the sol by using special stabilizers in order to maintain their individual stability and to prevent coagulation. Magnetorheological fluids are the suspensions of larger particles, one to three orders of magnitude larger than colloidal ferrofluid particles. The magnetorheological particles contain hundreds of magnetic domains. These liquids demonstrate dramatic changes in their rheological behavior in response to an externally applied magnetic field.

Magnetic micro- and nanospheres as well as monolith gels made of cross-linked polymers have been studied extensively for a wide range of applications. Polymer-encapsulated magnetic filler particles with diameters of less than 1 μm are of some interest in pharmaceuticals, cosmetics, as well as in paint production due to their improved physical and mechanical properties. Magnetic separation of labeled cells and other biological entities, therapeutic drug, gene and radionuclide delivery, radio frequency methods for the catabolism of tumors via hyperthermia, and contrast enhancement agents for magnetic resonance imaging applications are the most important examples [66].

The synthesis of stimuli-responsive polymer gel microspheres have been receiving growing attention [67–69, 73]. Polymer microspheres that combine both temperature- and pH-responsive volume phase transitions have been elaborated, as reported by many authors [2, 5–7, 65, 66]. Owing to their relatively rapid and easy magnetic separation, thermosensitive polymer magnetic microspheres could be widely used in biomedical and bioengineering, such as for enzyme immobilization and immunoassay, cell separation, and clinical diagnosis. In addition due to their sensitivities to both magnetic field and temperature, thermosensitive polymer magnetic microspheres offer a high potential application in the design of a targeting drug delivery system, which is considered a safe and effective way for tissue-specific release of drugs. With a small quantity of magnetic thermoresponsive polymer microspheres, a large amount of drug could be easily administered and transported to the site of choice.

3
Preparation of Magnetic Polymer Gels and Elastomers

Preparation of magnetic polymer gels and elastomers is similar to that of other filler-loaded networks. One can precipitate well-dispersed colloidal-sized particles in the polymeric material. The in situ precipitation can be made before, during, and after the cross-linking reaction [74, 88]. According to another method the preparation and characterization of colloidal magnetic particles are made separately, and the cross-linking takes place after mixing the polymer solution and the magnetic sol [75].

We have prepared monolith magnetic polymers as well as highly swollen magnetic polymer gels. It is worth distinguishing two kinds of filler-loaded samples:

- Isotropic composites, with uniform spatial distribution of filler particles
- Anisotropic composites, characterized by uniaxially ordered filler particles

Magnetic micro- and nanogel beads containing magnetic nanoparticles were also prepared and studied.

3.1
Preparation of Magnetic Polymer Gels with Uniform Filler Distribution

3.1.1
Magnetite-Loaded Poly(vinyl alcohol) Gels

Chemically cross-linked poly(vinyl alcohol) (PVA) hydrogel filled with magnetite particles were prepared as follows. First, magnetite (Fe_3O_4) sol (ferrofluid) was prepared by a conventional coprecipitation method. The sediment was dispersed with 1 M HCl or $HClO_4$, which induced peptization. The particle size was varied over a wide range from nanometers to microns by the concentration of the reactants as well as by the stirring intensity. Small angle X-ray scattering was used to determine the size and the size distribution of the magnetite particles in the ferrofluid.

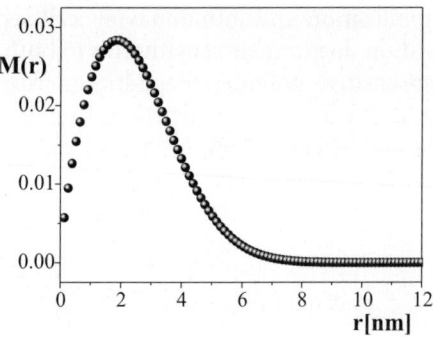

Fig. 2 Size distribution of the magnetite particles in the magnetite sol; $M(r)$ is mass fraction of particles of diameter r [76]

Figure 2 shows the size distribution of a representative sample. In this case the average diameter of the magnetite particles was found to be 4 nm.

In order to prepare magnetite-loaded PVA gels (mPVA), the stabilized magnetite sol having a concentration of 10 wt % was mixed with PVA solution. Poly(vinyl alcohol) (PVA 72 000, Merck-821038) is a neutral water-swollen polymer which reacts with GDA resulting in chemical cross-linkages between PVA chains. For the preparation of the magnetic PVA gels, 8 wt % PVA (Merck) solution and 1 M glutaraldehyde GDA (Aldrich) were used as the polymer and the cross-linker, respectively. HCl was used as initiator. The cross-linking density was varied by the amount of GDA relative to the vinyl alcohol VA units of PVA chains. The ratio of [VA] units to [GDA] molecules was varied between $100 \leq [VA]/[GDA] \leq 400$. Spherical, cylindrical, and cube shaped samples were prepared for mechanical and swelling measurements. After gelation, the samples were kept in distilled water to remove the unreacted compounds. A more detailed description of the preparation process can be found in one of our previous papers [29].

3.1.2
Magnetic Poly(dimethylsiloxane) Elastomers with Uniform Filler Distribution

For the preparation of magnetic field-responsive poly(dimethylsiloxane) composites (mPDMS), a commercial two-component reagent (*Elastosil 604 A* and *Elastosil 604 B*) provided by Wacker (Munich) was used. Component A contains polymers and the Pt-containing catalyst while component B provides the cross-linking agent. Component B was varied from 2.5 wt % to 3.5 wt %. Carbonyl-iron and iron oxide particles were used as magnetic filler particles. The concentration of the solid magnetic materials was varied between 10 wt % and 30 wt % in the polymer matrix. Figure 3 a shows a scanning electron microscope (SEM) picture of the carbonyl iron particles. It can be seen that polydisperse carbonyl iron particles have a spherical shape with

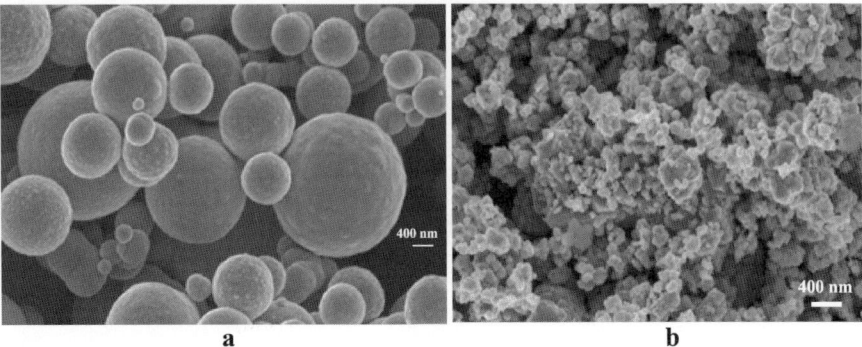

Fig. 3 SEM picture of the carbonyl iron (**a**) and iron oxide (**b**) particles

a smooth surface. The average diameter of these particles was found to be 2.5 μm. The magnetite (Fe_3O_4) particles (Fig. 3b) have smaller average size (0.2 μm) and narrower size distribution. It must be mentioned that the aggregated structure of magnetite particles as shown in Fig. 3b is due to SEM preparation. In solution as well as in the gels, the individual particles are distributed randomly as confirmed by small angle X-ray scattering.

Both carbonyl iron and iron oxide particles were dispersed in the *Elastosil 604 A*. After mixing it with the *Elastosil 604 B* component, the solution was transferred into a cube-shaped mould. The cross-linking reaction was carried out at ambient temperature for 4.5 h to obtain the magnetic composites. After cross-linking polymerization, the cubed, cylindrical, and spherical samples were removed from the moulds [77, 78].

3.2
Preparation of Anisotropic Polymer Composites

Synthesis of elastomers under uniform magnetic field can be used to prepare anisotropic samples.

The first step in the preparation is to mix the filler particles with the reaction mixture containing the polymers, the cross-linking agent, and the catalyst. The second step is to stabilize the system in order to avoid aggregation and sedimentation of the solid particles.

In order to prepare anisotropic elastomers, the reaction mixture in the mould was placed between the poles of a large electromagnet (JM–PE–I JEOL, Japan) (Fig. 4).

The mixture was subjected to $B = 400$ mT uniform magnetic field. The imposed field orients the magnetic dipoles and if the particles are spaced closely enough, mutual particle interactions occur. Due to the attractive forces, a pearl chain structure develops as shown in Fig. 5. This phenomenon is called the magneto-rheological effect [53, 55, 57].

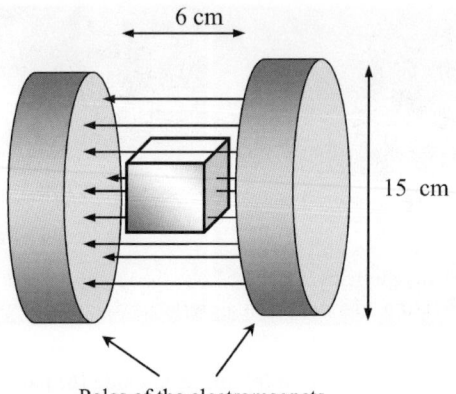

Poles of the electromagnets

Fig. 4 Preparation of uniaxially ordered polymer composite under uniform magnetic field

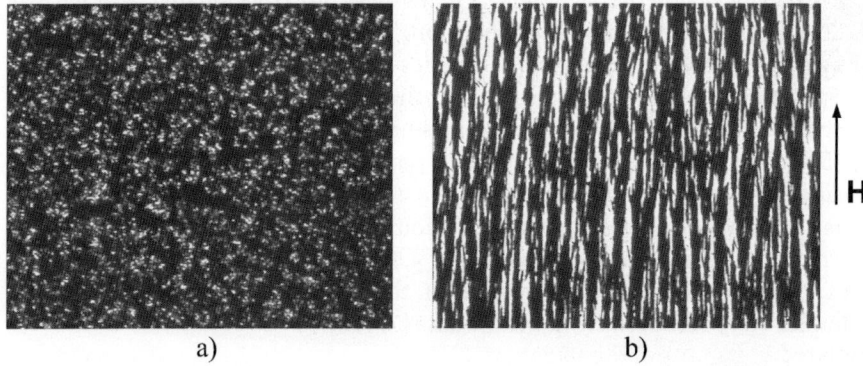

Fig. 5 Magnetorheological effect of carbonyl iron particles in uniform magnetic field. **a** Magnetite suspension in the absence of external field. **b** The same suspension in the presence of 100 mT uniform magnetic field. The direction of forming chains is parallel to the field direction, as shown by the *arrow* [78]

The columnar structure can be fixed by the cross-linking reaction. After the cross-linking polymerization was completed, the samples were removed from the moulds [77, 78].

Due to the magneto-rheological effect, the resulting magnetic composite becomes anisotropic in terms of mechanical and magnetic properties. One can easily vary the direction of the particle chains by the direction of the applied magnetic field, as shown in Fig. 6.

Depending on the concentration of the magnetic particles as well as on the applied magnetic field, columnar structures of the magnetic particles built in the elastic matrix can be varied over a wide range. Figure 7 shows a PDMS elastomer containing a dense column of pearl chains of magnetic particles.

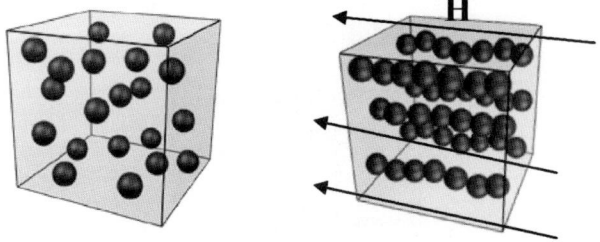

Fig. 6 Effect of uniform magnetic field on the structure formation of magnetic particles. *Arrows* indicate the direction of uniform magnetic field. *H* denotes the magnetic field intensity

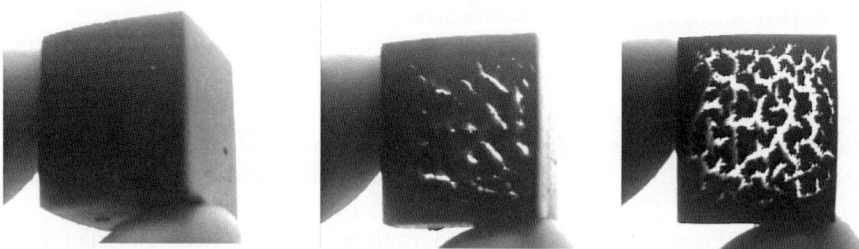

Fig. 7 Columnar structure of iron particles built in PDMS elastomer. In order to visualize the columnar structure of magnetic particles in the PDMS composite the sample is rotated. The angle of rotation has been decreased from *left* to *right*

On the basis of the same principle, anisotropic mPVA hydrogels were also prepared using magnetite nanoparticles.

3.3
Preparation of Magnetic Gel Beads

3.3.1
Preparation of Magnetic Polystyrene Latex

In order to prepare magnetic polystyrene (mPS) latex, the miniemulsion polymerization method was applied [79, 80]. A water-soluble surfactant (sodium lauryl sulphate, SLS) and a monomer-soluble long-chain alcohol, the so-called costabilizer (stearyl alcohol, SA), were used for stabilizing the styrene/water miniemulsion against both coalescence (with SLS) and diffusional degradation or Ostwald ripening (with SA), respectively. The magnetite nanoparticles in a ferrofluid (surface-treated magnetite nanoparticles in toluene) were introduced into the monomer under stirring to obtain the stable magnetite/styrene dispersion. Preparation of the miniemul-

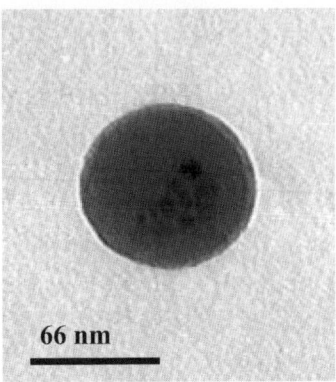

Fig. 8 TEM micrograph of a mPS microsphere

sion was carried out by ultrasound homogenization. An oil-soluble initiator N,N'-azobis(isobutyronitrile) (AIBN) was used for starting the polymerization reaction. The detailed polymerization procedure can be found in [80].

Figure 8 shows a typical TEM micrograph of a mPS particle. The magnetite particles are visible as dark spots inside the polystyrene bead. The encapsulation of the magnetic iron oxide in the polymer was found to be complete; neither free magnetite particles nor free polystyrene latex particles not containing magnetite were observed. This observation is in agreement with the concept suggested by Landfester [79] that droplet nucleation is the preferred mechanism and, thus, every single droplet is nucleated for an ideal miniemulsion system. Although the magnetite particles were distributed uniformly in the starting styrene system, they aggregated and accumulated at one side of the spheres during the microemulsion preparation and the subsequent polymerization.

3.3.2
Preparation of Millimeter-Sized PNIPA and Magnetic mPNIPA Thermosensitive Gel Beads

Among stimuli-responsive or thermoresponsive polymer microspheres, poly(N-isopropylacrylamide) (PNIPA) was investigated most intensively. Cross-linked PNIPA is a thermosensitive hydrogel which undergoes a volume phase transition at its lower critical solution temperature (LCST) of around 34 °C. Several authors reported the synthesis of PNIPA microspheres [68–73].

Chemically cross-linked temperature sensitive poly(N-isopropyl acrylamide) gel beads were prepared from NIPA, methylenebisacrylamide cross-linker (BA), ammonium persulphate initiator (APS), and N,N,N',N'-tetramethylenediamine (TEMED) from Aldrich Chemicals. These chemicals were used without further purification.

We have prepared gel beads with an average diameter of 2.0 mm and narrow size distribution using the drop method. The chemical procedure was based on the method developed by Park and Choi [81]. An interpenetrated network (IPN) was prepared by simultaneous gelation of Ca-alginate, with a concomitant free radical polymerization of NIPA and cross-linker within the beads. Alginate was dissolved in 25 mL of 0.01 M Tris buffer (1.75% w/v, degassed and then mixed with 1.92 g NIPA monomer, 0.09 g BA (4% to NIPA w/v), 0.125 mL TEMED, and 0.89 mL of ferrofluid. The above solution was added dropwise using a syringe needle into 300 mL of 0.01 M Tris buffer solution, containing 3% $CaCl_2$ and 0.1% (w/v) APS. Gel beads were kept for about 30 min under nitrogen atmosphere. At the end of the polymerization (24 h), they were washed three times with distilled water to remove unreacted monomer, cross-linker, and initiators. Afterwards, the beads were equilibrated with 0.1 M EDTA solution (pH $\sim$ 7) for 2–3 h to chelate calcium ions and extract the alginate from the IPN beads.

3.3.3
Preparation of Magnetic Polystyrene Latex Covered by PNIPA Gel Shell

The magnetic polystyrene latex particles with diameter of 120 nm were covered by PNIPA gel layer. The mPS latex prepared according to the procedure described previously was strongly stirred at 60 °C and kept under N_2 atmosphere for 1 h. Then, 0.05 g APS and 0.5 mL 1 M NIPA solution were added to the mPS latex and the reaction mixture was stirred at 60 °C for more than 1 h. Then, 0.5 mL 1 M NIPA solution and 0.36 mL 0.1 M BA solution were added. After 2 h, 0.5 mL 1 M NIPA solution and 0.36 mL 0.1 M BA solution were added again to the mixture. This mixture was stirred 60 °C for more than 2 h under N_2 atmosphere. Figure 9 shows the structure of the core-shell microsphere in dry state.

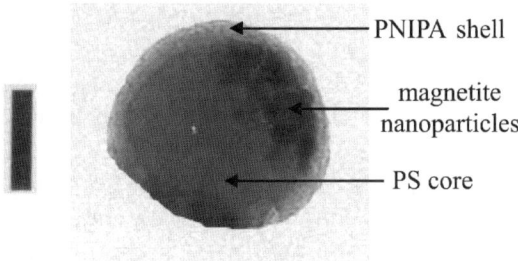

Fig. 9 TEM photograph of mPS-PNIPA core-shell latex in dry state. The *bar* denotes 66 nm [82]

4
Magnetic Properties of Ferrogels

In the magnetic polymer gels (ferrogels), the finely distributed nanosized magnetic particles are located in the swelling liquid and are attached to the flexible network chains by adhesive forces. The solid nanoparticles are the elementary carriers of a magnetic moment. If the concentration of magnetic particles is below the percolation threshold, the gel contains a collection of single domain particles. From the magnetic viewpoint, the ferrogel is a dilute ensemble of non-interacting magnetic moments. Similarly to ferrofluids, at elevated temperatures the ferrogel should show superparamagnetic characteristics. But in contrast to magnetic fluids, the position of the particles built into the gel are rigidly fixed and thus the moments of the blocked particles cannot rotate towards the direction of the external field if the Zeeman energy is smaller than the anisotropy energy (this is the situation in fields smaller than 0.5 T).

In the absence of an external magnetic field, the moments are randomly oriented, and thus the gel has no net magnetization. As soon as an external field is applied, the magnetic moments tend to align with the field and produce a bulk magnetization, M [53, 55]:

$$M = \frac{\langle \mathbf{M} \cdot \mathbf{H} \rangle}{H} = M_S \langle \cos\theta \rangle \,, \tag{1}$$

where θ is the angle between the magnetic field and the magnetic axis of a particle, M_S is the saturation magnetization, H is the external field, and the average $\langle \cos\theta \rangle$ is taken over all the particles. With ordinary field strengths, the tendency of the magnetic moments to align with the applied field is partially overcome by thermal agitation, such as with the molecules of paramagnetic gas. As the strength of the field increases, all the particles eventually align their moments along the direction of the field, and as a result, the magnetization saturates. The single domain particles are therefore the most efficient magnets.

If the field is turned off, the magnetic dipole moments begin to randomize and thus the bulk magnetization reduces. Depending on the temperature and size of particles there are two mechanisms for the relaxation of magnetization [83]:

i) Brownian rotation, where the rotation of the magnetic moments takes place together with the rotation of the individual particles with a relaxation time, τ_B

ii) Néel relaxation, which is a rotation of the magnetic moments inside the particles; the Néel mechanism can be described by another relaxation time, τ_N.

In ferrogels, the magnetic particles are trapped by the polymer network, therefore the Brownian rotation is supposed to be restricted and the Néel relaxation is expected to become more effective.

On the basis of the relaxation times, one can define a blocking temperature, T_B. At temperature $T < T_B$ the magnetic material exhibits ferromagnetic characteristics with hysteresis and remanent magnetization. With increasing temperature the ferromagnetic characteristics decrease and vanish at the blocking temperature. Above T_B, all apparent ferromagnetic characteristics disappear, even though within each particle the magnetic moments remain ferromagnetically aligned. It is worth mentioning that the blocking temperature is proportional to the volume of the magnetic particles, which means even a modest increase in particle size can result in a significant increase in the value of T_B. At $T > T_B$, the magnetic nanoparticles exhibit superparamagnetic behavior with a giant magnetic moment, which is about 10^4 times larger than the individual atomic moment.

The magnetic behavior of superparamagnetic material can be described by the Langevin function. Assuming the magnetization of individual particles in the gel to be equal to the saturation magnetization of the pure ferromagnetic material, the magnetization, M, of ferrogel in the presence of an applied field can be expressed as:

$$M = \Phi_m M_s L(\xi) = \Phi_m M_s \left(\coth \xi - \frac{1}{\xi} \right), \qquad (2)$$

where ϕ_m stands for the volume fraction of the magnetic particles in the whole gel, and the parameter, ξ of Langevin function $L(\xi)$ is defined as:

$$\xi = \frac{mH}{k_B T}, \qquad (3)$$

where μ_0 is the magnetic permeability of the vacuum, m is the giant magnetic moment of nanosized magnetic particles, k_B denotes the Boltzmann constant, and T stands for the temperature.

According to Eq. 2, the magnetization of a ferrogel is directly proportional to the concentration of magnetic particles and their saturation magnetization. It is worth mentioning that the Langevin approach can be used either:

- If the particles are rigidly fixed in the gel and the energy barriers between the easy axes for spin alignment are smaller than the thermal energy
- The direction of the magnetization within the particle is fixed, but the rotation of particles in the gel is allowed

The static properties are in either case the same, but the relaxation times are different. On the basis of Eq. 2, one may conclude that two experimental features should characterize superparamagnetism:

- There is no hysteresis in the field dependence of the magnetization (M is a single valued function of H)
- The magnetization is an universal function of $\frac{H}{T}$

Figure 10 shows a typical magnetization curve of ferrogel. The magnetization is plotted against the magnetic field strength. Four cycles have been measured and plotted.

On the basis of this figure, we can conclude that within the experimental accuracy no hysteresis loop has been observed. This finding says that no remanent magnetization takes place in ferrogels at room temperature. This is an important result, which means that in alternating magnetic field the transformation of magnetic energy into thermal energy is rather small. For this reason, devices made of ferrogels and subjected to alternating magnetic field are characterized by small energy loss per cycle. It may also be seen that the

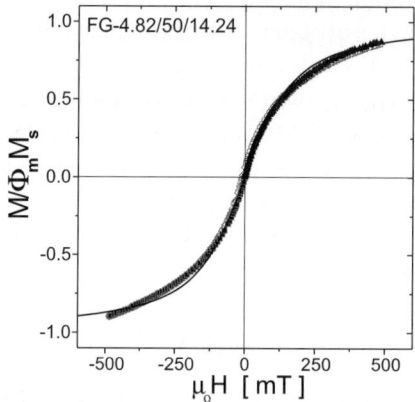

Fig. 10 Magnetization curve of a ferrogel measured at room temperature. The *solid line* was fitted by Eq. 2 with parameters $M/\phi_m M_s = (\coth \xi - 1/\xi)$, where $\xi = 0.0145 \mu_0 H$ and $\Phi_m M_s = 2.33$. *Different symbols* represent subsequent cycles

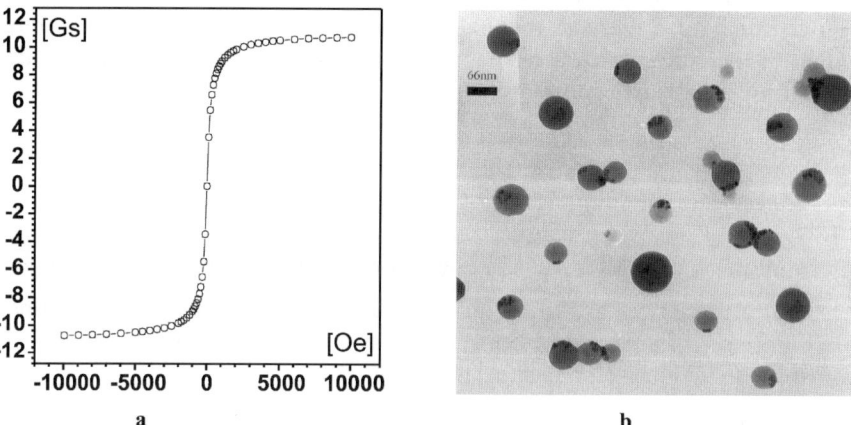

Fig. 11 Magnetization curve (**a**) and TEM image (**b**) of mPS latex. The magnetization is expressed in Gauss and the fields intensity is given in Oersted

shape of $M(H)$ dependence can be satisfactorily described by the Langevin approach, shown by the solid line in the figure.

Similar magnetic behavior has been found for mPS microspheres. Figure 11a shows a typical magnetization curve for mPS particles, whose TEM is shown in Fig. 11b.

The mPS microspheres show a typical superparamagnetic behavior at room temperature. No remanence was observed when the magnetic field was removed.

5
Elastic Properties of Magnetic Composites Studied by Unidirectional Compression

In the absence of an external magnetic field, the magnetic gels and polymer networks present a mechanical behavior very close to that of a swollen filler-loaded network. Since a typical magnetic gel can be considered as a dilute magnetic system $\phi_m < 0.1$, we may neglect the influence of magnetic interactions on the modulus. Thus the stress–strain dependence of a unidirectionally deformed gel sample can be expressed on the basis of Gaussian statistical theories [3, 4, 84].

Consider a circular cylindrical gel extended along its axis so that the circular symmetry is maintained. Let h_0 and h be the initial (reference) and deformed length of the cylinder. Then, the principal deformation ratio along the axis z is λ_z; λ_x and λ_y are deformation ratios for the axes perpendicular to the axis z. The strain energy density function for and ideal phantom network, W_{el}, can be expressed as follows [3, 4, 84]:

$$W_{el} = \frac{1}{2}G\left(\lambda_x^2 + \lambda_y^2 + \lambda_z^2 - 3\right), \tag{4}$$

where G represents the elastic modulus. If the deformation is realized at constant volume, $\lambda_x \lambda_y \lambda_z = 1$, then λ_x and λ_y reduce to $\lambda_x = \lambda_y = \lambda_z^{-\frac{1}{2}}$. The strained energy can be expressed in terms of the G and the deformation ratio:

$$W_{el}(\lambda_z) = \frac{1}{2}G\left(\lambda_z^2 + \frac{2}{\lambda_z} - 3\right). \tag{5}$$

The nominal stress, σ_n, can be derived from Eq. 5 by a standard method [84–86]:

$$\sigma_n = \left(\frac{\partial W_{el}}{\partial \lambda_z}\right). \tag{6}$$

The result is neo-Hokean stress–strain dependence:

$$\sigma_n = G\left(\lambda_z - \lambda_z^{-2}\right) = GD, \tag{7}$$

where $D = (\lambda_z - \lambda_z^{-2})$. The elastic modulus of magnetic composite can be expressed as a function of filler concentration:

$$G = G_0 f(\phi_m), \tag{8}$$

where G_0 denotes the modulus of gel without colloidal filler particles and $f(\phi_m) > 1$ describes the reinforcement effect due to the filler–polymer interactions [87]. The modulus of a ferrogel can be varied by the cross-linking density through G_0 and by the concentration of colloidal particles via $f(\phi_m)$.

In order to characterize the deviation from the ideal phantom network behavior, the Mooney–Rivlin representation is often used:

$$\frac{\sigma_n}{D} = C_1 + C_2 \lambda_z^{-1}, \tag{9}$$

where C_1 and C_2 are the Mooney–Rivlin parameters. C_2 is usually consider as the measure of non-ideality [84–86].

We have performed unidirectional stress–strain measurements to characterize the elastic properties of the magnetic composites. The static and dynamic mechanical tester (INSTRON 5543) was used for the experiment. Both large deformation and small deformation behavior were investigated. In order to determine the elastic modulus, we studied the small strain compression of magnetic composites. All the mechanical measurements were carried out at room temperature. The elastic modulus of the magnetic samples containing randomly distributed magnetic materials was determined on the basis of Eq. 7 from the plot of nominal stress against D. The slope, which provides the elastic modulus G, was calculated by the linear least squares method. Figure 12 shows the dependence of nominal stress on the quantity D for different samples containing randomly distributed carbonyl iron particles. The Mooney–Rivlin representation is also presented in Fig. 12, left.

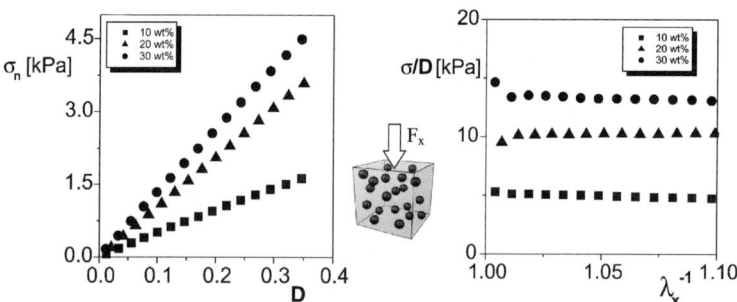

Fig. 12 Result of mechanical measurement performed on a magnetoelast filled with randomly distributed carbonyl iron. The concentrations of the filler particles are indicated on the figure. The cross-linker content was 3 wt % in each case. Neo-Hookean (*left figure*) and Mooney–Rivlin representations (*right figure*) are plotted

It is seen in the figures that the magnetoelast shows ideal mechanical behavior in the studied deformation range. Similar ideal mechanical behavior was observed for other magnetoelasts characterized by different cross-linking densities and different amount of fillers dispersed randomly in the elastic matrix. It has to be mentioned that within the experimental accuracy (5%) no hysteresis has been found.

5.1
Anisotropic Mechanical Behavior

In Sect. 3.2 the preparation of magnetic polymer composites under uniform magnetic field has been described. The resulting composites show anisotropic behavior. The anisotropy manifests itself in the direction-dependent elastic modulus. Figure 13 shows carbonyl iron-loaded mPDMS elastomers. All three samples contain the same amount of filler particles, but the spatial distribution of the filler is different, as shown in the figure.

The experimental data were analyzed on the basis of Eq. 7. It is seen that the slopes of the straight lines (elastic moduli, G) are direction-dependent. The elastic modulus is larger if the compression force and the direction of pearl chain structure are parallel. This finding indicates a strong mechanical anisotropy. It can be concluded that the spatial distribution of the solid particles has a decisive effect on the stress–strain dependence.

When the direction of the compressive force is parallel and perpendicular to the pearl chain structure, a deviation has been found from the ideal mechanical behavior. The nominal stress does not obey a linear dependence with the quantity D, as demonstrated in Fig. 14. This kind of mechanical behavior can be described by the Mooney–Rivlin equation with $C_2 < 0$.

More detailed analysis of experimental data can be found in our previous paper [78]. In contradiction to the isotropic sample, the stress–strain depen-

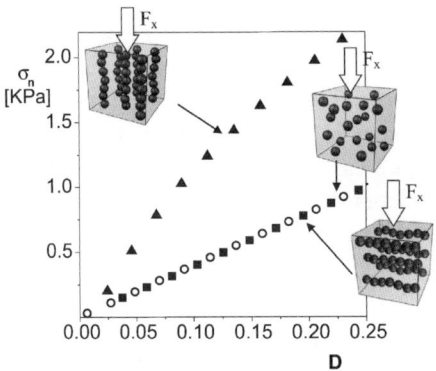

Fig. 13 Anisotropic mechanical behavior of a mPDMS sample containing 30 wt % magnetite. The *arrows* on the samples indicate the direction of compressive force

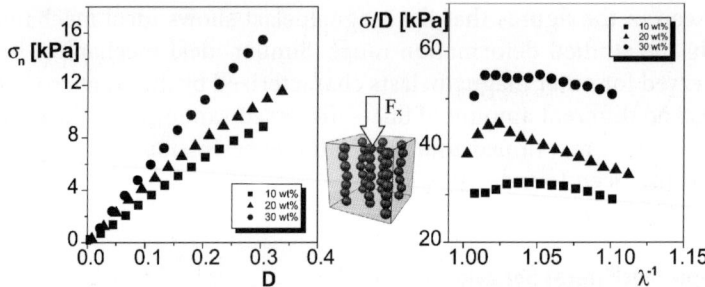

Fig. 14 Deviation from the ideal mechanical behavior for anisotropic magnetoelasts having different amounts of filler particles, as indicated in the figures. The cross-linker content was 3.5 wt % in every case

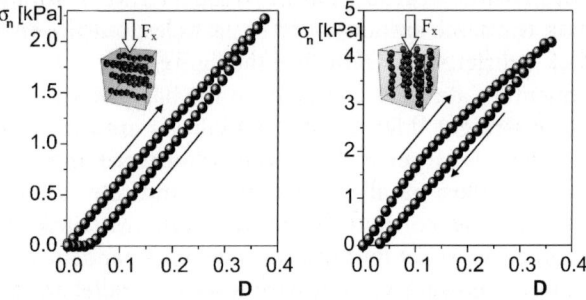

Fig. 15 Hysteresis phenomena of the anisotropic magnetoelasts. The *arrows* on the figures indicate the compression and the elongation of the sample. Carbonyl iron-loaded samples (2.5 wt % cross-linker, 30 wt % carbonyl iron)

dence of the anisotropic samples exhibits hysteresis. This is demonstrated in Fig. 15. An interpretation of hysteresis phenomenon is given in Sect. 5.2.

5.2
Anisotropic mPVA Gels and mPDMS Networks at Large Deformation

Stress–strain dependence for **mPVA** gels as well as for **mPDMS** samples with random and aligned structures is shown in Fig. 16. When comparing the shape of the curves one can find significant differences. Not only the slope (elastic modulus, G) depends on the filler structure locked in the elastic materials, but also the shape of the curves is significantly different. The field structured composite along the chain-like structure exhibits a larger modulus. One can also see a characteristic change in the slope of the stress–strain curve of the anisotropic samples when the direction of the compression is parallel to the pearl chain structure. The stress–strain curve for **mPVA** gel filled with magnetite shows a bending point (Fig. 16a) if the compression and particle alignment are parallel. The critical strain at which the bending occurs

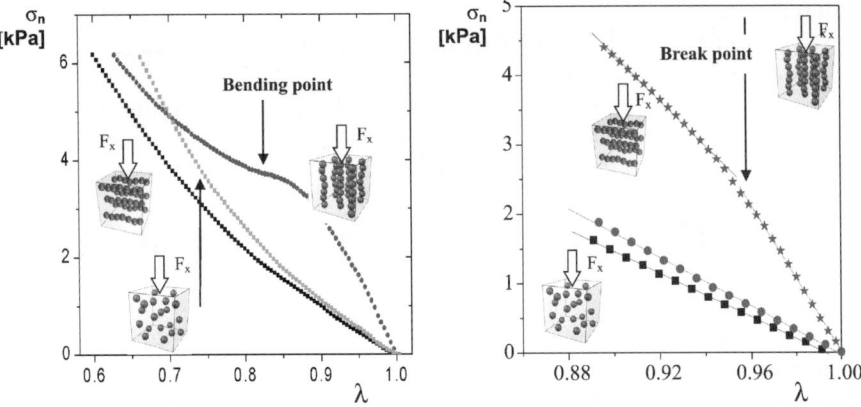

Fig. 16 Typical stress–strain behavior of mPVA (**a**) and mPDMS (**b**) gels in unidirectional compression measurements. In both cases, three samples having the same amount of filler particles are compared, but the distribution of the particles is different

was found to be $\lambda_c = 0.85$. At larger deformation $\lambda < \lambda_c$ the slope (elastic modulus) decreases, which means a stress-induced softening phenomenon. The softening may be interpreted as a mechanical instability due to the presence of rod-like particle chains. The behavior of a rod subjected to longitudinal compressing force was first investigated by Euler. It was found that at small compressing force the rods are stable with respect to any small perturbations. This means that the rods are slightly bent by some compressive force and the sample tends to return to its original shape when the force ceases to act.

If the compressive force exceeds a critical value, f_c, a mechanical instability occurs. This results in a large bending. The behavior of bending above the critical strain, λ_c, requires less force ($f < f_c$) for further deformation. The softening phenomenon is illustrated in Fig. 17.

When the deformation ratio reaches the value 0.9, the pearl chain structure starts to bend under compression, because the polymer chains in the mPVA network interact with the magnetic particles, as shown in Fig. 17. The

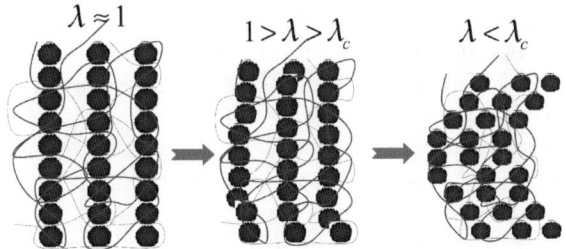

Fig. 17 Schematic representation of the bending of the magnetic PVA gels under compression

ordered structure of the particles and the interaction with the polymer network prevent the pearl chain structure of the particles from being destroyed by compression.

The deformation parallel to the structure has a strong influence on the mechanical behavior because of the bending of the ordered structure. The anisotropic magnetite-loaded PDMS samples showed similar behavior under compression.

In contrast to the magnetic mPVA gels, compression of the carbonyl iron-loaded mPDMS network results in a break-point in the stress–strain curve if the deformation of the sample is parallel to the pearl chain structure. Figure 16b shows that the nominal stress increases with the compression up to a deformation ratio of 0.95 in every case. On increasing the compression above this ratio, the columnar structures of the iron particles are destroyed (Fig. 18).

In a uniform magnetic field, the iron particles interact with each other, but the particles do not interact with the polymer network. The polymer network can immobilize the ordered structure, but it cannot prevent the break-up of the ordered structure under compression. The iron particles in the chains slip out from the columnar structure, breaking the ordered structure and resulting in hysteresis, as shown in Fig. 15.

In order to characterize the bending or breaking phenomena we have defined an apparent elastic modulus (G_a) as follows:

$$G_a = \frac{\partial \sigma_n}{\partial D}. \tag{10}$$

Figure 19 shows the dependence of the apparent elastic modulus on the quantity D. For the sake of comparison, results of compression measurements performed on both isotropic and anisotropic samples are shown. The composites contain Fe_3O_4 filler particles with concentration of 30 wt %. In the case of isotropic samples, the apparent elastic modulus increases slightly within the experimental error of 5%. For anisotropic samples, the apparent modulus increases significantly under deformation up to the value $D = 0.85$. Above this value, the modulus does not change notably.

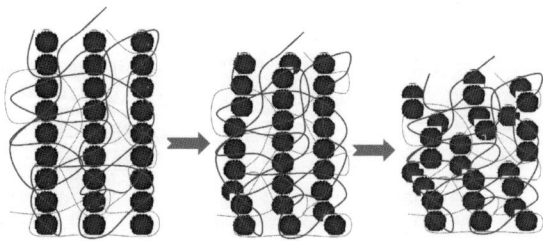

Fig. 18 Schematic representation of structural change of the iron-loaded mPDMS elastomer under compression

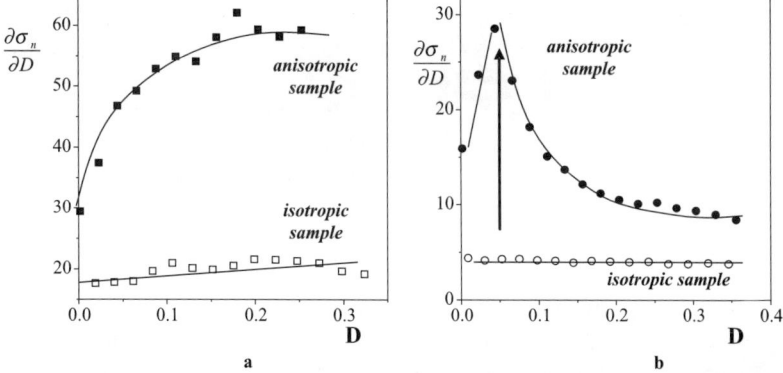

Fig. 19 Change in the apparent elastic modulus due to compression: **a** Fe_3O_4-loaded mPDMS sample, **b** iron-loaded mPDMS sample. *Solid lines* are guides for the eyes and the *arrow* indicates the breaking point [77]

Figure 19b shows the results of **mPDMS** composites filled with carbonyl iron. The concentration of filler particles is 30 wt %. One can see the breaking of the pearl chain structure at $\lambda = 0.95$.

It is also seen that the change of the apparent elastic modulus of the isotropic sample is negligible. For the anisotropic samples, the apparent elastic modulus increases significantly under deformation up to the value 0.05 D. Above this value, the modulus decreases notably, which can be explained by the broken structure.

6
Effect of Uniform Magnetic Field on the Elastic Modulus

It has only recently been accepted that the elastic properties of magnetic elastomers can be increased rapidly and continuously by application of an external magnetic field [11–14, 98–101].

If the magnetoelast contains magnetic particles dispersed randomly, there are two basic experimental situations: The compressive force (F_x) and the direction of magnetic field (characterized by the magnetic induction, B) can be either parallel or perpendicular, as shown on Fig. 20. In one of our previous papers, we found a 20% increase of the modulus for a magnetoelast containing randomly distributed magnetic particles [37].

As shown in Sect. 3.2, elastic materials with tailor-made anisotropy can be prepared under an external field. Jolly et al. have found 0.6 MPa maximum increases in the shear modulus for iron-loaded elastomer [17]. In this work, the shear elastic behavior perpendicular to the columnar structure has been investigated. To the best of our knowledge, no information is available concerning other experimental situations. Depending on the direction of the

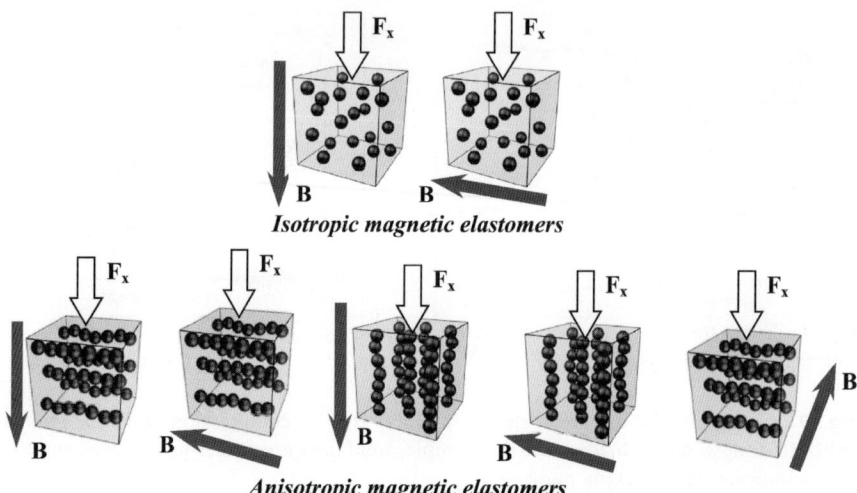

Fig. 20 Experimental possibilities in the study of the influence of external magnetic field on the elastic modulus. *White arrows* indicate the direction of the force; *black arrows* show the direction of the magnetic field

magnetic field, the columnar structure as well as the mechanical stress, the elastic modulus, can be determined by using five different experimental setups (see Fig. 20).

6.1
Mechanical Measurements Under Static Magnetic Field

The elastic modulus of magnetic composites was measured under uniform magnetic field at 293 K using two different homemade instruments. The magnetic induction, B, was measured by a Phywe Teslameter. Comparing the direction of the force and the magnetic field, there are two different situations: the force and the magnetic field are either perpendicular or parallel as shown in Fig. 21a,b. In the perpendicular case, the magnetic polymer is placed on the sample stage between two magnetic poles (Fig. 21a) and the stress, which is perpendicular to the magnetic field, is measured by moving the silver rod connected to the strain gauge. The gap between the magnetic poles is 40 mm and the cross section of the pole is 19.6 cm^2. The magnetic field around the sample stage can be considered to be uniform. We were able to measure the stress–strain dependence under a magnetic field of 0–400 mT.

When the stress and the strain are parallel (Fig. 21b), a coil is used to create the magnetic field. In this case, we were able to measure the stress–strain dependence at 0–100 mT. In the experiments, the magnetic field intensity was varied and the elastic modulus was measured as a function of magnetic induction, B.

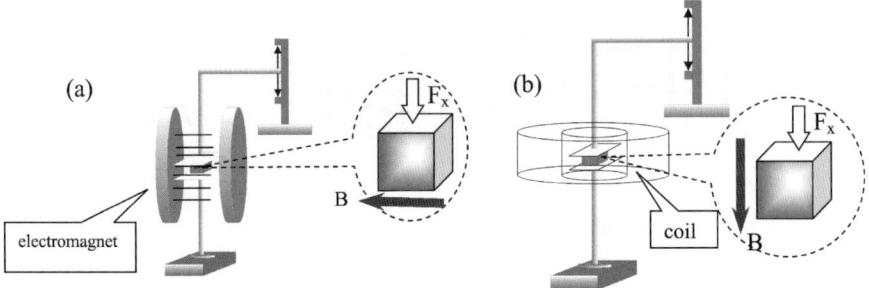

Fig. 21 Schematic representation of the experimental set-up measuring elasticity under uniform magnetic field. F_x represents the force and B denotes the magnetic induction

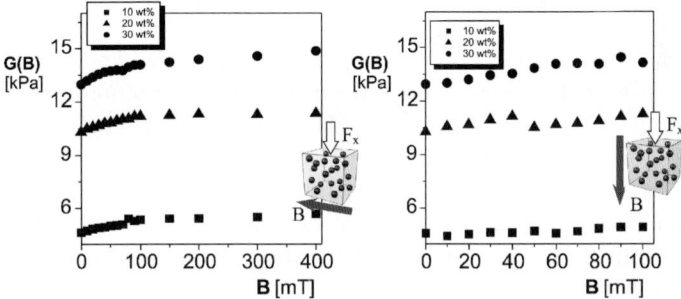

Fig. 22 Dependence of the magnetic field intensity on the elastic modulus for mPDMS containing different amounts of randomly distributed magnetite particles. The concentrations of the filler particles are indicated in the figure

Figure 22 shows the effect of uniform magnetic field on the modulus of mPDMS samples containing randomly distributed magnetite particles. Depending on the direction of the field to the mechanical stress, two kinds of experimental arrangements were used: parallel and perpendicular. One can see in both figures that there is a slight increase in the modulus due to the external field. This finding is in accordance with our previous result where similar temporary reinforcement effect was reported for magnetite (Fe_3O_4)-loaded poly(vinyl alcohol) hydrogels [37]. It is also seen that the concentration of the filler particles increases the field-free modulus, G_0. Larger amount of magnetite results in a larger elastic modulus.

6.2
Compressive Force is Perpendicular to the Direction of Particle Chains

The effect of the uniform magnetic field on the elastic modulus was studied when the applied mechanical stress was perpendicular to the particle align-

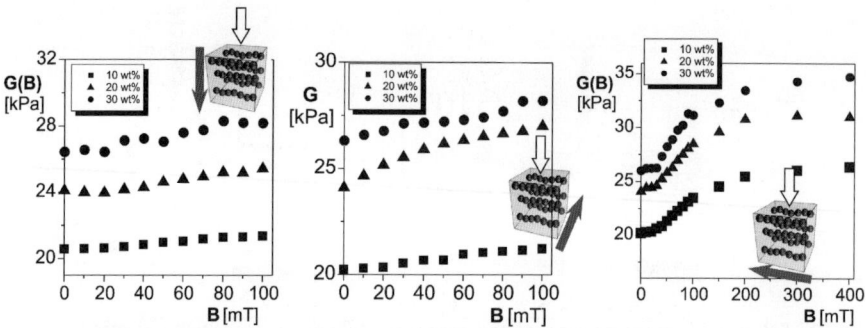

Fig. 23 Effect of the magnetic field intensity on the elastic modulus. The iron content of the elastomers is indicated in the figure. The *white* and *black arrows* show the direction of the force and the uniform magnetic field, respectively

ment. By varying the direction of the applied magnetic field and the columnar structure of the particles, we have three possible experimental arrangements: the direction of the field is perpendicular to the particle chains (Fig. 23 left and center) or parallel to the particle chains (Fig. 23 right).

On the basis of the experimental results, the effect of the uniform magnetic field on the elastic modulus is shown in Fig. 23. It may be concluded that the spatial orientation of the force, the field, and the particle arrangement play a decisive role in the temporary reinforcement effect. A weak effect has been found when the field is perpendicular to the particle alignment (Fig. 23 left and center).

If the columnar arrangements of the particles are parallel to the direction of the magnetic field, the elastic modulus increases significantly (Fig. 23 right). At small field intensities of up to 30 mT, a slight increase has been observed. Above 30 mT, the modulus increases significantly. At higher field intensities (from 200 mT), the elastic modulus tends to level off. It is also seen that by increasing the concentration of the iron particles in the polymer matrix the elastic modulus, G_o, increases. The most significant reinforcement effect was found if the magnetic field is parallel to the particle alignment.

6.3
Compressive Force is Parallel to the Direction of Particle Chains

Figure 24 shows the temporary reinforcement effect when the direction of the applied mechanical stress is parallel to the particle alignment. In this case, there are two different experimental situations. The direction of the magnetic field is parallel or perpendicular to the direction of the columnar structure as seen in Fig. 24. When the applied uniform magnetic field is perpendicular to the direction of the particle chains (Fig. 24b), the magnetic field does not

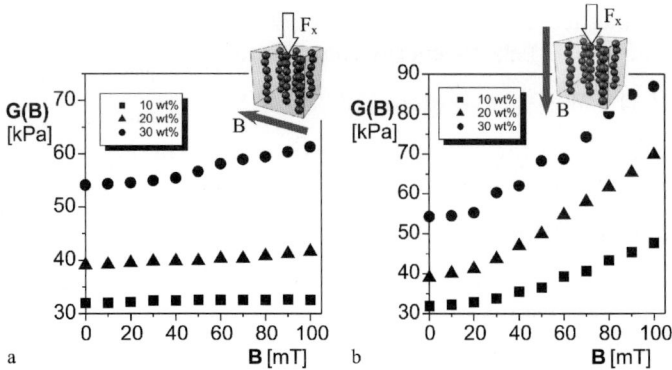

Fig. 24 Dependence of the magnetic field intensity on the elastic modulus. The arrangements of the particles in the polymer networks are parallel to the applied mechanical stress while the applied uniform magnetic field is parallel (**a**) or perpendicular (**b**) to the columnar structure. The concentration of the carbonyl iron particles in the mPDMS matrix is indicated

make its significant influence on the elastic modulus. The strongest magnetic reinforcement effect was found when the applied uniform magnetic field was parallel to the particle alignment and to the mechanical stress (Fig. 24b). The relative increment of the modulus had the highest value.

When the iron content is increased, the mutual particle interaction increases, which induces higher excess modulus. The elastic modulus increases significantly with the magnetic field intensity if the particle arrangements are parallel to the force and to the applied uniform magnetic field. However, the magnetic field does not influence the elastic modulus significantly when the applied field is perpendicular to the mechanical force and to the particle chains.

6.4
Magnetic Elastomers in Uniform Field. A Theoretical Approach

As it follows from Eq. 7, the elastic modulus can be defined as follows:

$$G_0 = \frac{1}{3} \lim_{\lambda_z \to 1} \left(\frac{\partial \sigma_{el}}{\partial \lambda_z} \right), \tag{11}$$

where G_0 represents the elastic modulus in lack of external magnetic field. Comparing Eqs. 6 and 11, one can express the elastic modulus by the following alternative expression:

$$G_0 = \frac{1}{3} \lim_{\lambda_z \to 1} \left(\frac{\partial^2 W_{el}}{\partial \lambda_z^2} \right). \tag{12}$$

In case of a magnetoelast, the overall energy density is the sum of $W_M(H_{\text{eff}})$ magnetic and $W_{\text{el}}(\lambda_z)$ elastic energy contributions:

$$W(\lambda_z, H_{\text{eff}}) = W_{\text{el}}(\lambda_z) + W_M(H_{\text{eff}}), \tag{13}$$

where H_{eff} stands for the effective magnetic field strength. The magnetic energy density, W_M of a piece of magnetoelast is related to the effective field:

$$W_M = \mu_0 M(H_{\text{eff}}) H_{\text{eff}}, \tag{14}$$

where $M(H_{\text{eff}})$ function describes the magnetization curve and μ_0 denotes the magnetic permeability of the vacuum.

Deformation of a magnetoelast involves a change in the particle arrangement, which affects the magnetic energy density, $W_M(H_{\text{eff}})$. As a consequence, the deformation of magnetoelast requires more energy in an uniform magnetic field since one must overcome both the change in the elastic, and the magnetic energy as well. In practice, this magnetic effect manifests itself as an increase in the elastic modulus.

On the basis of Eqs. 12 and 13, the elastic modulus measured under uniform external field can be expressed as the sum of two contributions:

$$G = G_0 + G_M^E, \tag{15}$$

where G_M^E denotes the magnetically induced excess modulus. We can characterize the magnetic field intensity by the magnetic induction, B, which is the product of the magnetic permeability of the vacuum and the field intensity: $B = \mu_0 H_{\text{eff}}$. The magnetic contribution of the modulus can be calculated from the change of magnetic energy due to the distortion of the composite. Similar to Eq. 11, the excess modulus can be defined as follows:

$$G_M^E = \frac{1}{3} \lim_{\lambda_z \to 1} \left(\frac{\partial^2 W_M}{\partial \lambda_z^2} \right). \tag{16}$$

Since the magnetic polarization is a fast process, consequently G_M^E develops quickly when the external field is turned on and disappears immediately when the external field is turned off. The analytic calculation of the magnetic excess modulus, G_M^E is a rather complicated task due to the fact that the magnetic energy density, $W_M(H_{\text{eff}})$, cannot be given in an analytical form. It was found that at low field intensities the magnetic excess modulus is proportional to the square of the magnetic induction: $G_M^E \propto B^2$. At high field intensities the magnetization saturates and as a result G_M^E approaches a maximum value of $G_{M,\infty}$.

These two limiting cases can be phenomenologically described by the following equation:

$$G_M^E(B) = G_{M,\infty} \frac{B^2}{a_B + B^2}, \tag{17}$$

where a_B represents a material parameter.

The parameters (a_B and $G_{M,\infty}$) can be determined by linearization of Eq. 16:

$$\frac{B^2}{G_M^E(B)} = \frac{a_B}{G_{M,\infty}} + \frac{1}{G_{M,\infty}} B^2 .\qquad(18)$$

By plotting the quantity of $B^2/G_M^E(B)$ against B^2, the slope provides the $1/G_\infty$ value, and the intercept gives the ratio $a_B/G_{M,\infty}$. We have found that the proposed phenomenological equation is in rather good agreement with the experimental results. Figure 23 shows the experimental data plotted according to Eq. 18.

We determined the parameters $G_{M,\infty}$ and a_B (Table 1) by the linear least squares method for the samples shown in Fig. 25.

In Fig. 26, the experimental data and the phenomenological approach are presented. It is seen that the agreement between experimental data and phenomenological description is quite satisfactory. Further, it is an important task to investigate how the parameters of Eq. 16 depend on the particle con-

Table 1 Parameters of Eq. 17 for the sample shown in Fig. 25, determined by the linear least squares method

Iron content [wt%]		a_B [(mT^2)]	$G_{M,\infty}$ [kPa]
10		12 059.55	6.68
20		6945.07	7.33
30		10 915.83	9.35

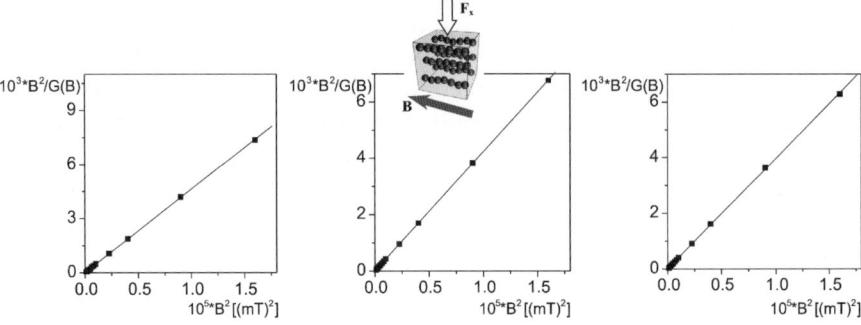

Fig. 25 Dependence of $B^2/G_M^E(B)$ on B^2 for mPDMS samples. The concentration of carbonyl iron varies from *left* to *right*: 10 wt%, 20 wt%, and 30 wt%. The experimental arrangement is shown in the figure

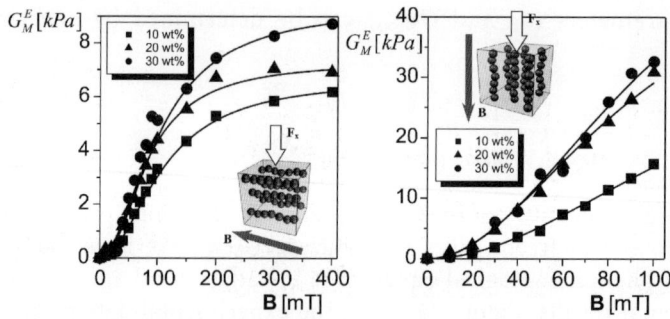

Fig. 26 Dependence of G_M^E on the magnetic induction for two kinds of magnetoelasts having different experimental arrangements. *Solid lines* were calculated on the basis of Eq. 17

centration as well as on the spatial distribution. This will be the subject of a further study.

The measure of magnetically induced excess modulus, G_M^E depends on the concentration and spatial distribution of the magnetic particles as well as on the strength of applied field. For randomly dispersed magnetic particles, G_M^E slightly depends on the particle concentration and on the magnetic induction. It was found to vary between zero and 2.8 kPa. A more significant magnetic reinforcement effect was found for anisotropic samples containing oriented particle chains, instead of randomly distributed particles. If the mechanical stress and the direction of columnar structure as well as the magnetic induction are all parallel, G_M^E approaches 32.7 kPa when the magnetic mPDMS sample contains 30 wt % carbonyl iron (Fig. 26).

7
Non-uniform Magnetic Field-Induced Deformation

When a magnetic composite is placed into a spatially non-uniform magnetic field, forces act on the magnetic particles, and the magnetic interactions are enhanced. The stronger field attracts the particles and, due to their small size and strong interactions with molecules of dispersing liquid and polymer chains, they all move together.

Changes in molecular conformation can accumulate and lead to shape changes (Fig. 27). The magnetic field drives and controls the displacement, and the final shape is set by the balance of magnetic and elastic interactions. The force density, f_m, on a piece of magnetic composite can be written as:

$$f_m = \mu_0(M\nabla)H , \qquad (19)$$

where μ_0 is the magnetic permeability of vacuum, M represents the magnetization, and ∇H takes into account the gradient of magnetic field, H. It should

 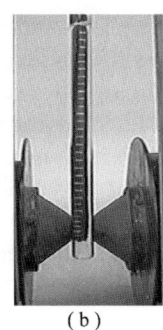 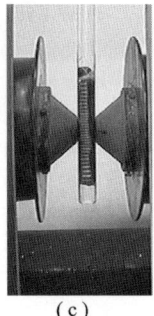 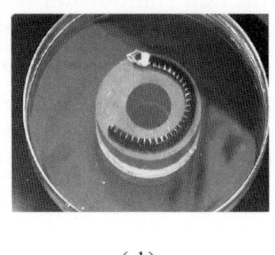

(a)　　　　　　(b)　　　　　　(c)　　　　　　(d)

Fig. 27 Shape distortion of ferrogels due to non-uniform magnetic field. **a** No external magnetic field. **b** The maximal field strength is located under the lower end of the gel. **c** The maximal field strength is focused in the middle of the gel along its axis, bending induced by a permanent magnet

be kept in mind that the magnetic force density vector varies from point to point in accordance with the position-dependence of product $(M\nabla)H$. The orientation of f_m is parallel to the direction of magnetic field. In a non-accelerating system, the force density manifests itself as a stress distribution, which must be balanced by the network elasticity. A completely balanced set of forces is in this respect equivalent to no external force at all. However, they affect the gel internally, tending to change its shape or size or both.

In general, the deformation induced by a magnetic field cannot be considered homogeneous, since the driving force $(M\nabla)H$ varies from point to point in space. However, one can find a special distribution of magnetic field, where the deviation from the homogeneous case is not significant. In this case, the condition for uniaxial deformation of a ferrogel cylinder can be written as follows [28–33]:

$$\lambda_z^3 - \beta(H_h^2 - H_m^2)\lambda_z - 1 = 0, \qquad (20)$$

where λ_H denotes the deformation ratio due to field induced strain. The parameter β is defined as:

$$\beta = \frac{\mu_0 \chi}{2G}, \qquad (21)$$

where χ stands for the initial susceptibility of the magnetoelast, H_h and H_m represent the magnetic field strength at the bottom and the top of a suspended ferrogel cylinder, respectively. Equation 19 can be considered as a basic equation for describing the unidirectional magnetoelastic properties. It says: if we suspend a magnetic composite in a non-homogeneous magnetic field in such a way that $H_h > H_m$, then elongation occurs. In the opposite case, when the field intensity is higher on top of the gel, i.e., $H_h < H_m$, Eq. 20 predicts unidirectional compression. The magnetic field-sensitive polymer gels can be made to bend and straighten, as well as to elongate and contract repeatedly many times without damaging the gel. The response time to obtain the new equilibrium shape was found to be less than a second and it seems to be independent of the size of the gel. It must be mentioned that all the shape changes reported here are completely reversible. Since the magnetic field can be created by electromagnets, it is easy to achieve dynamical conditions by modulated current intensity. We applied stepwise and sine-wave modulation by a function generator in the frequency range of 0.01–100 Hz. A cylindrically shaped magnetoelast characterized by a height of 8 mm and radius of 4.5 mm was put onto the upper surface of a standing electromagnet [36] as shown in Fig. 28.

Figure 28 shows that the magnetoelastic response time is rather short, one cycle requires half second. We have to mention that up to 40 Hz the magnetic stimulus and the elastic response are strongly coupled. Neither phase shift, nor significant mechanical (or magnetic) relaxation takes place.

We have studied the dependence of elongation on the steady current intensity required by the electromagnets to produce an external magnetic field. A cylindrical ferrogel was suspended in water to prevent evaporation of swelling liquid and to balance the weight of gels by the buoyancy. The experimental arrangement is shown in Fig. 29.

Fig. 28 Snapshot of shape change of a magnetoelast due to modulated magnetic field. The frequency of the field is 40 Hz

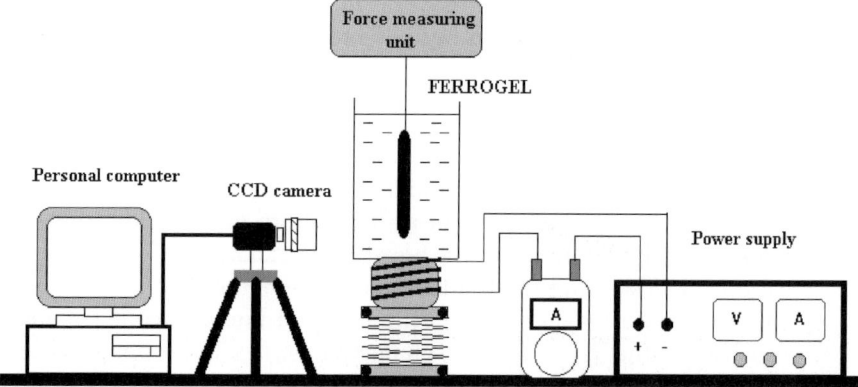

Fig. 29 Experimental arrangement to measure the stress and strain induced by a magnetic field

We have also studied the magneto-elastic properties of ferrogel cylinders suspended in water vertically between plan-parallel poles of electromagnets, as shown in Fig. 30.

Figure 31a shows the effect of magnetic field on the deformation of magnetic elastomer. The relative displacement is plotted against the steady cur-

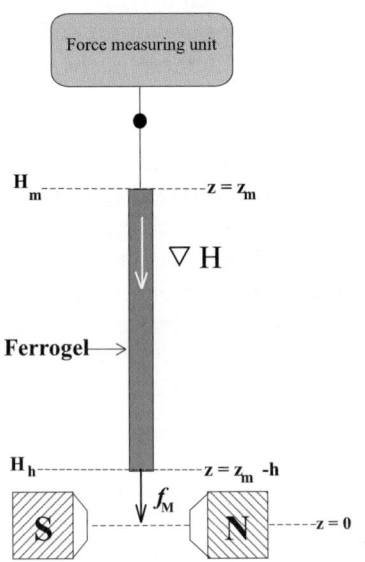

Fig. 30 Schematic diagram of experimental setup to study the magnetoelastic properties

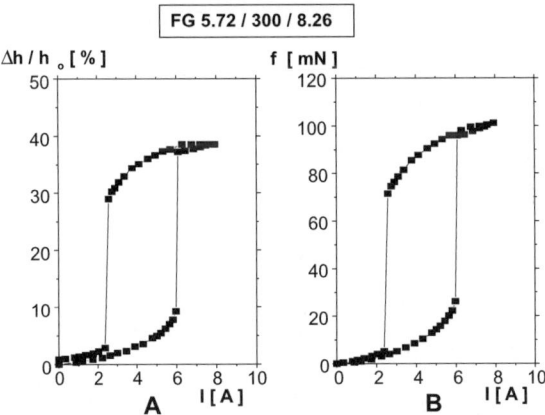

Fig. 31 The non-continuous magneto-elastic behavior of a ferrogel. **a** Relative displacement as a function of steady current, **b** Force induced by magnetic field as a function of steady current. The initial length of the gel was $h_0 = 163$ mm. The magnetic field was induced by a steady current flowing through the face-to-face plan-parallel magnetic poles of the electromagnet

rent intensity, which is proportional to the maximal field intensity. It can be seen that the displacement of the lower end of the ferrogel – due to magnetic force – is rather significant. A giant magnetostriction takes place. In many cases, we were able to produce an elongation of 40% of the initial length by applying a non-uniform magnetic field. It may be seen that at small current intensities the displacement slightly increases. However, at a certain current intensity a comparatively large, abrupt elongation occurs. This non-continuous change in size appears within an infinitesimal change in the steady current intensity. Further increase in the current intensity results in another small extension.

We have found that by decreasing the current a contraction takes place. Similarly to the extension, the measure of the contraction was found to have a non-continuous dependence on current intensity. It is worth mentioning that the discrete shape transition occurred within a time interval of one second, independently of the gel size. Not only the relative displacement, but also the measured force, show similar dependence characterized by both an abrupt change in magnitude and hysteresis. This is demonstrated in Fig. 31b.

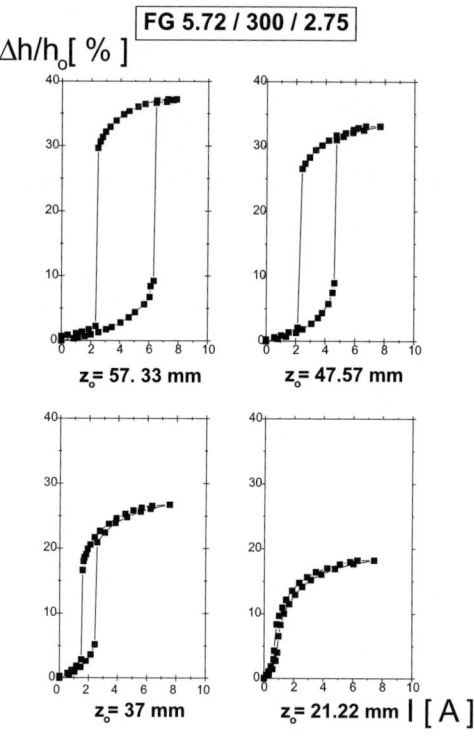

Fig. 32 Effect of initial position on the discontinuous shape transition, $h_0 = 163$ mm. Values of z_0 are indicated in the figure

The peculiar property of ferrogel magneto-elasticity is the hysteresis that characterizes the extension–contraction process. Despite its irreversible nature, this process is time-independent under ordinary conditions. It must be emphasized that the observed hysteresis phenomena is not a consequence of the well-known magnetic hysteresis of ferromagnetic materials, since according to our measurements the ferrogels exhibit no magnetic hysteresis at all (Fig. 31).

By variation of the experimental conditions, we have found a cross-over between continuous and discontinuous shape transitions. The initial position in the non-uniform magnetic field seems to play an essential role in the mode of stretching (Fig. 32). The cross-over between continuous and non-continuous transitions seems to be determined by the position of gel only. Similar observation has been made for other ferrogels having much smaller amount of magnetite. We have varied the magnetite concentration of ferrogels in wide range between 2.75 and 12.6 wt %. For other soft magnetic composites similar abrupt shape transition have been observed.

7.1
Interpretation of Non-continuous Shape of Transition

It is possible to interpret the abrupt transformation and hysteresis phenomena on the basis of Eq. 20. In order to find the dependence of elongation on the steady current intensity, first we have to relate the magnetic field strengths, H_h and H_m to the steady current intensity. Let us assume that the magnetic field strength varies along the gel axis as:

$$H(z) = H_{max}h(z) , \qquad (22)$$

where H_{max} represents the maximal field strength at the position $z = 0$ and $h(z)$ is a unique function characterizing the experimental arrangement (geometry of poles and gap distance).

It was found that the z-directional distribution of magnetic field strength can be satisfactorily approximated by the following forms:

$$h(z) = \begin{cases} 1 - kz^2 & \text{if } |z| < \delta \\ (1 - k\delta^2)e^{-\gamma(|z|-\delta)} & \text{if } |z| \geq \delta \end{cases} , \qquad (23)$$

where γ is a characteristic constant describing the exponential decay of field strength at larger distances, δ is the radius of poles, and the constant $k = \frac{\gamma}{2\delta + \gamma \delta^2}$ was determined by taking into account the same slope of $h(z)$ curves at distance $r = \delta$, where the functions (in Eq. 22) approach each other. According to the Biot–Savar law, H_{max} can be written as $H_{max} = k_I I$, where k_I is a proportionality factor and I is the steady current intensity. It must be mentioned that the value of parameter k_I strongly depends on the quality of the electromagnet. The numerical solution of Eq. 19 provides the λ dependence, which is shown in Fig. 33.

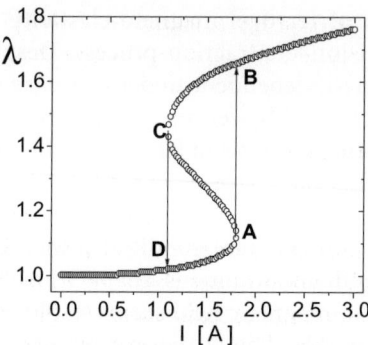

Fig. 33 Dependence of deformation ratio, λ, on the steady current intensity as calculated on the basis of Eqs. 20–23. For the calculation $h_0 = 10$ cm, $z_0 = 5$ cm, and $\gamma = 0.4$ were used

In contradistinction to the state of equilibrium, the internal parameter (deformation ratio) is not a single-valued function of the external parameter (current intensity), i.e., metastability develops with abrupt non-equilibrium transition. The transition from A→B takes place at a different value of current intensity from that for C→D. An area ABCD remains a kind of hysteresis loop. This cycle depicted in Fig. 33 is time-independent and may be repeated several times. Here, a macroscopic energy barrier is provided by magnetic interactions and this energy barrier compels the ferrogel to go around it a higher or lower current than the equilibrium value.

We have also studied the influence of parameters z_0 on the shape of $\lambda(I)$ dependence. These results are summarized in Fig. 34. One can see that with increasing z_0 both the measure of abrupt transition and hysteresis increases. This finding is in accordance with our experimental results (see Fig. 33).

The parameter γ, which controls the gradient of the field strength, also plays an essential role. It was found that the necessary condition for realiz-

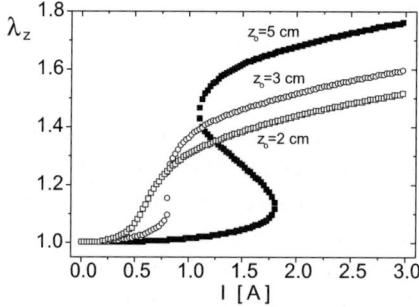

Fig. 34 Dependence of deformation ratio, λ_z, on the steady current at different initial position of the gel

ing an abrupt transition requires a value, of $\gamma > 0.15$. Below this value no discontinuous transition occurs if $h_0 = 10$ cm and $z_0 = 5$ cm.

7.2
Non-homogeneous Deformation

Description of the macroscopic deformations of ferrogels requires a special treatment due to the complex nature of the mechanism of magnetic field-induced deformations. The non-linear character of both elastic and magnetic interactions results in some novel features of the deformation process. However, not only the high degree of non-linearity of the governing equations, but also the non-homogeneity of the resulting deformations makes the treatment special and fairly complicated. As a body force, the magnetic field induces deformations with a unique pattern never seen in deformations induced by surface tractions.

In the conventional theory of elasticity [84–86], a linear relationship between stress and strain is assumed. In this way, the elastic nature of a material is fully described by a tensor of the elastic moduli. For most solids in ordinary situations, this approximation is acceptable since the deformations are small. Most of the magnetism-related elastic effects, such as magnetostriction can be treated within the frame of the linear theory. The reason for this is that materials with significant magnetic susceptibility (i.e., ferro-, antiferro-, and ferromagnetic materials) are hard solids and, consequently, are unable to sustain large deformations even in fields as high as several Tesla. In ferrogels, however, superparamagnetism provides a fair magnetic susceptibility and the extremely high elasticity of the polymer network results in a material capable of undergo giant deformations (the strain can reach 1.5) in an ordinary magnetic field. It is clear that for ferrogels we have to employ the generalized, non-linear theory in order to properly describe the deformation process in the whole deformation range.

In the theory of non-linear elasticity, a material body is regarded as a set of elastically joined material points. The forces are considered to act on the material points and the condition of mechanical equilibrium is characterized by the balance between external and elastic forces. External forces are classified by means of their nature as surface tractions and body forces. Surface tractions act only on surface points, while body forces act on each material point. Naturally, body forces can be generated only by fields, of which the most common is gravity which acts universally and homogeneously on each material element. Although at first glance the treatment of gravity in the theory of elasticity seems quite straightforward, nevertheless, the resulting deformation is always non-homogeneous, i.e., the displacement of material points during the deformation varies point by point. The reason for the non-homogeneity is that the stress changes vertically, analogous to hydrostatic pressure. As two popular examples of gravity-induced deformations, which can be found in any elasticity textbook, we refer to the bending of a stick and the distortion of a standing column under their own weight. The remaining two fields, namely electric and magnetic fields, can induce even more interesting deformation patterns. Unlike gravity, these fields might have a complex distribution in space. Since both the field and the force are non-uniform, different material points experience force of different strength and direction, which leads to a non-homogeneous deformation, often with a complex deformation pattern.

In order to point out the essential difference between deformations induced by body forces and surface tractions we recall Ericksen's theorem [88]. The theorem states that homogeneous deformations are the only deformations that can be achieved by the application of surface tractions alone, considering a homogeneous and isotropic material characterized by an arbitrary strain–energy function. In other words, we cannot induce diverse non-homogeneous deformations with surface tractions. In contrast to surface tractions, application of fields that act as body forces leads to non-homogeneous deformations without any additional constraints for the material.

Not only the nature of the resulting deformation, but also the treatment of body forces is different from that of surface tractions. Many well-established and successful methods for non-linear elasticity (e.g., inverse method, diagonalization of the deformation tensor) are not applicable due to the non-homogeneity of the deformation. In [43], we discuss how to describe magnetic field-induced non-linear deformations of a hyperelastic, incompressible magnetic continuum characterized by Langevin-type magnetization and a neo-Hookean strain–energy function. The latter is frequently used for swollen polymer gels. Although we are concentrating on the deformations of ferrogels, the presented theory is general in a sense that it is also applicable for electric field-induced deformations.

As we point out in [42], due to the relatively complex form of the Langevin-type magnetization and the magnetic force density – even in simple magnetic field distributions – it is not possible to achieve an analytical solution of the driving equations. In order to demonstrate the characteristics of the magnetic field-induced deformations of a magnetic continuum, it is worth examining a simple one-dimensional situation.

Let us consider a very simple physical situation similar to our unidirectional experiments discussed in the previous section. Namely, a long and thin ferrogel cylinder is suspended in water vertically. The magnetic field is induced by a solenoid-based electromagnet placed under the gel. The axis of the gel cylinder (z) is parallel to the magnetic field and its gradient. In this case, the deformation of the gel is uniaxial and can be considered to be one-dimensional:

The governing equation for this situation describing the displacement of each point of the gel along the z axis is the following second-order, non-linear ordinary differential equation [42]:

$$G\left(\frac{d^2 u_z(Z)}{dZ^2} + \frac{2}{(du_z(Z)/dZ)^3}\frac{d^2 u_z(Z)}{dZ^2}\right) + M[u_z(Z)]\frac{dH[u_z(Z)]}{dZ} = 0, \quad (24)$$

where $u_z(Z)$ represents the displacement relative to the reference (undeformed) configuration, G is the shear modulus of the gel, M denotes the magnetization, and H stands for the magnetic field strength. As boundary conditions, the displacements and/or the surface tractions must be prescribed on the two ends of the gel cylinder. In our particular case, the position of the top surface of the gel cylinder was fixed by a rigid, non-magnetic copper thread, while the bottom surface was free and unloaded. Accordingly:

$$u_z(0) = 0, \quad t(Z_m) = 0, \quad (25)$$

where 0 and Z_m represent the position of the top and bottom end of the gel, respectively.

Based on Eq. 23 with boundary conditions given by Eq. 24, we calculated the unidirectional deformation of a ferrogel cylinder. This is shown in Fig. 35. On the left-hand side, the magnetic field strength along the z axis is plotted. The distribution of the field we employed in the calculations was similar to that in real experiments. As one can see, the gel elongates as the magnetic field intensity increases. At a certain field intensity, the gel falls abruptly into a new equilibrium position, similar to that observed experimentally. The white lines on the gel body demonstrate the non-homogeneity of the deformation. Different distances between adjacent lines indicate different degrees of deformation. The high degree of non-homogeneity is clearly

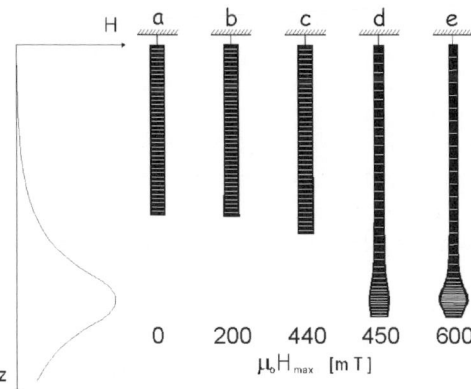

Fig. 35 Schematic representation of the uniaxial deformation of a ferrogel cylinder calculated numerically from Eq. 24. The external magnetic field distribution is shown on the *left*. Gel *a* is undeformed ($B = 0$). Gels *c* and *d* represent the abrupt transition within a slight increase of the field intensity

seen. At lower field intensities (gels *b* and *c*), the upper part of the gel elongates to a greater extent than the lower part, whereas at higher field intensities (gels *d* and *e*) the lower part of the gel contracts while the upper part elongates.

In order to test the validity of our model, we compared theoretical calculations with the results of unidirectional experiments. In Fig. 36, the strain of the bottom end of the gel is plotted against the maximum field intensity. Both experimental and calculated points are shown. The calculated points fit quite well the measured ones, indicating that our model is able to reproduce not only the non-continuous characteristic of the deformation process, but also provides accurate, realistic numerical values.

As mentioned previously, the analytical determination of the displacement field is generally extremely complex, therefore a numerical procedure is necessary to calculate the deformation. Nowadays the finite element method (FEM) is prevalent in solid mechanics and provides a useful tool for studying realistic, three-dimensional non-linear deformations. However, magnetic body forces are not commonly applied in engineering, consequently they cannot be simply considered with the usual FEM systems. Some FEM programs allow us to extend the program with user-defined subroutines for modeling various material behaviors, load types, or other special effects. The FEM system MARC has been chosen for the calculations because it was developed for modeling non-linear mechanical deformations and can be extended with the aforementioned user subroutines. As an illustrative example of three-

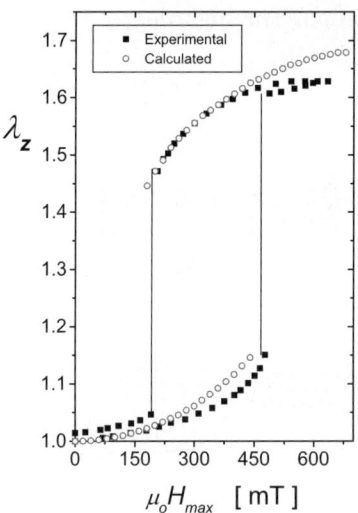

Fig. 36 Uniaxial elongation of a ferrogel cylinder. The *points* represent the displacement of the bottom end of the gel. The *blank points* were calculated on the basis of Eq. 24 with boundary conditions and magnetic field distribution given by Eqs. 22 and 23, respectively

dimensional deformation of ferrogels, in Fig. 37 we present the deformation of a ferrogel block whose ends are fixed to a wall.

The magnetic field is uniaxial and has a Gaussian distribution along the gel axis, as shown on the top of the figure. The arrows indicate the location of the maximum field intensity. We represented the deformation of the block at different field intensities. Looking at the set of pictures, one may associate the changes with the motion of a living worm. This indicates that the deformation and motion of ferrogels bears a close relation to that of simple living organ-

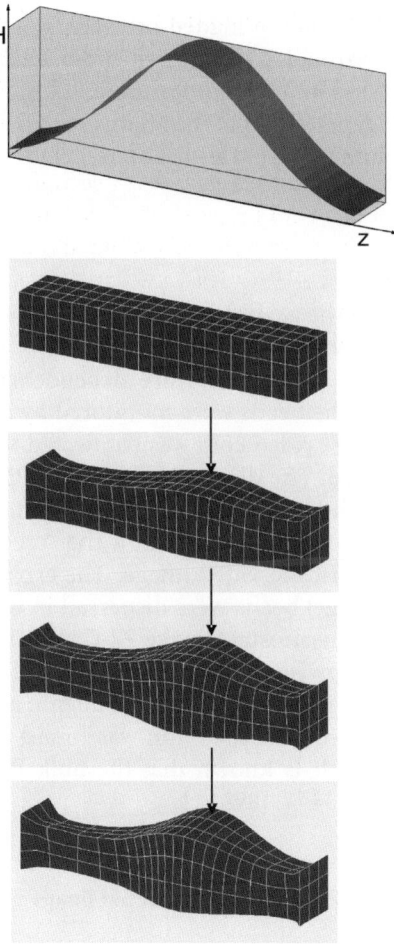

Fig. 37 FEM calculation of the deformation of a ferrogel block in a uniaxial magnetic field at different field intensities. The magnetic field strength has a Gaussian distribution along the block as shown in the figure. The position of the field maximum is indicated by *arrows*

isms. This is due to the complexity and non-homogeneity of the magnetic field-induced deformations.

8
Effect of the Magnetic Particles on the Swelling Behavior of PNIPA and PDMS Gels

The influence of magnetic nanoparticles on the collapse transition as well as on the swelling and shrinking kinetics is the subject of this investigation. Gel beads of millimeter size and monolith gels of centimeter size were prepared from both PNIPA and magnetite-loaded mPNIPA gels. Swelling and shrinking behavior was studied and compared in order to establish the influence of magnetic nanoparticles on the equilibrium and kinetic behavior of PNIPA gels. The temperature dependence of the relative swelling degree, q_r, was used to characterize the volume phase transition:

$$q_r(T) = \frac{r_T}{r_{10}}, \tag{26}$$

where r_{10} is the radius of the gel bead at 10 °C, and r_T represents the radius of the gel bead at T arbitrary temperature.

In order to determine the temperature dependence of the swelling degree, volume changes of the beads were monitored by a digital video system. A CCD camera with a 1/3″ video chip was connected to a PC through a real-time video digitalizer card. The diameter, $d = 2r_T$ of PNIPA and mPNIPA gel beads was followed on the magnified picture by home-developed software. Small changes in the diameter (one pixel on the screen) can be monitored and measured on the real-time video image. The error of the measurement was within 0.01 mm. The gel beads were dispersed in water and the temperature was controlled by a thermostat (Haake P2-C30P). The temperature of the system was increased stepwise from 10 °C to 50 °C. The increment of temperature was 2 °C and the incubation time was 20 min at each temperature. In the case of millimeter-sized beads, 20 min was found to be enough to reach the swelling equilibrium. It is known that the bulk PNIPA gel has a phase transition temperature at 34 °C [89–91].

8.1
Temperature Sensitivity of PNIPA and mPNIPA Gel Beads

Figure 38 shows the dependence of relative swelling degree on the temperature for PNIPA and magnetite-loaded mPNIPA gels. Both kinds of gels have the same polymer concentration and cross-linking ratio. The only difference is the presence of magnetite.

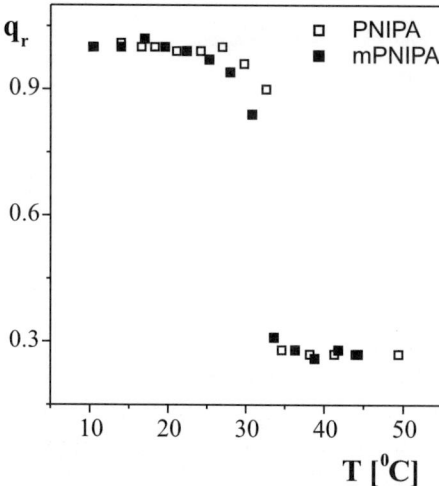

Fig. 38 Temperature dependence of the relative swelling degree of PNIPA and mPNIPA gels. The molar ratio of the monomer and the cross-linker molecules [NIPA]/[BA] is 50. The magnetite content of mPNIPA gel is 2.1 wt %. The initial diameter of the gel beads was 1 mm

One can see in the figure that for both gel systems the temperature dependence of volume change is not continuous; an abrupt change occurs when the temperature exceeds 30 °C. It is also seen that within the experimental accuracy no difference was observed between PNIPA and mPNIPA gel beads. The presence of magnetic nanoparticles influences neither the measure of volume change nor the collapse transition temperature (abbreviated as T_C). A careful analysis based on derivatives of the q_r–T curves has shown that for both kinds of PNIPA beads, T_C was found to be 32 °C. We have also studied the effect of cross-linking density on the volume phase transition. Figure 39 shows these results.

It can be concluded that the cross-linking density does not make its influence felt on the temperature dependence of relative swelling degree. It is worth mentioning that the volume change exceeds 300%. The same effect was found for magnetite-loaded PNIPA gels as well. In contrast to this behavior, the temperature dependence of the mass swelling degree is influenced by the cross-linking degree of PNIPA gels.

Shrinking kinetics on PNIPA and mPNIPA gel disks were investigated and compared. Figure 40 shows these results. It can be seen that by decreasing the cross-linking density, the relative swelling degree increases as expected and significant difference in the kinetics of unloaded and magnetite-loaded PNIPA disks can be observed. This difference is due to the formation of a surface skin layer on the unloaded PNIPA gels.

This skin layer blocks the water molecules from leaving the polymer matrix, resulting in bubbles on the surface. The presence of the magnetite

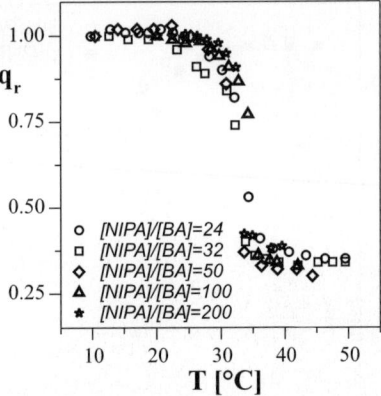

Fig. 39 Effect of the cross-linking density on the temperature-dependent relative swelling degree of PNIPA beads. The cross-linking ratio is indicated on the figure. The initial diameter of the NIPA gel beads was 1 mm

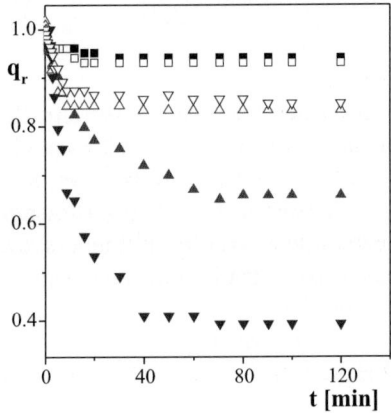

Fig. 40 Time dependence of the relative swelling degree of PNIPA and mPNIPA gel discs. The volume change is due to heating from 20 to 40 °C. The initial diameter and thickness of the gel discs was 5 mm and 2 mm, respectively. The molar ratios of the monomer and the cross-linker molecules ([NIPA]/[BA]) in NIPA gels are: 50 (□), 100 (o), 200 (◊), and in MNIPA gels are: 50 (■), 100 (•), 200 (♦)

nanoparticles modifies the structure of the polymer network and the surface properties. This modified surface retards the formation of the surface skin layer and bubbles as shown in Fig. 41.

In contradistinction to the PNIPA gel disks, the gel beads of millimeter size have no skin layer and bubbles on the surface. The lack of skin layer and bubbles can be explained by the structure of the gel beads. For the preparation of the beads the interpenetration network formation was used. This method results in two polymer networks within each other. If one is removed from the

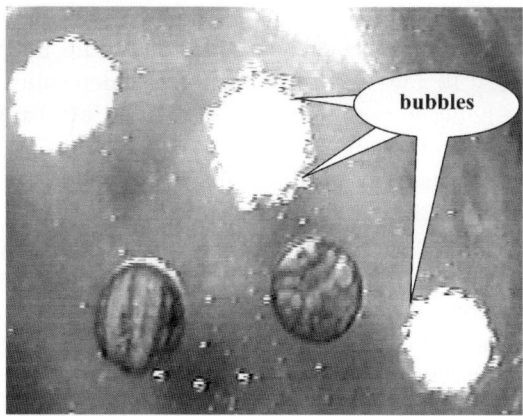

Fig. 41 Surface skin layer formed on PNIPA gel discs at 40 °C. The *dark discs* represent the magnetite-loaded mPNIPA gel disks

other, a new polymer matrix with channels can be obtained. These channels guide the water molecules out of the gel beads during the phase transition, preventing skin formation.

Shrinking kinetics measurements were also performed in order to study the influence of the surface skin layer. The kinetics of shrinking of poly(N-isopropylacryalmide) and magnetic poly(N-isopropylacryalmide) gels are significantly different. The presence of magnetic nanoparticles retards the formation of a surface skin layer and bubble formation and results in faster shrinking kinetics.

8.2
Effect of the Magnetic Particles on Swelling Kinetics of mPDMS Elastomers

Spherical mPDMS gels containing chain-like filler particles parallel to each other are good candidates for checking the influence of the direction-dependent elastic modulus on the swelling kinetics. On the basis of the kinetics of swelling [92–97] one expects a slower volume relaxation in the direction parallel to the columnar structure, and a faster kinetics in the perpendicular direction. As a consequence, the spherical beads become anisotropic during the swelling and the kinetics can be characterized by the direction-dependent relaxation times.

In order to investigate the direction-dependent relaxation time of isotropic as well as anisotropic gels, swelling kinetics experiments were performed. Filler-loaded spherical mPDMS samples with an initial diameter of 2 cm were used for the experiment. The gel samples were placed into cyclohexane, and the diameter of the gel was recorded by a digital Sony video camera (DCR-TRV40E).

Due to the long swelling time (15 h), 2 s in every 5 min were recorded. Then the movies were digitalized (by a Pinnacle Studio interface card) and one picture at every 10 s was captured for the size determination. Thus, the elapsed real time between two pictures was 25 min. MS Paint software was used for the size determination.

In order to determinate the swelling kinetics, two kinds of diameter were measured as a function of time, as shown in Fig. 42.

It follows from the thermodynamics of swelling that the higher the modulus, the smaller the equilibrium swelling degree. Since our magnetite-loaded samples show a strong direction-dependent elastic modulus, after swelling the increment of the size due to the volume change will not be the same.

This finding is evidenced by Fig. 43. The anisotropic filler-loaded mPDMS sample was put into cyclohexane, which is a good solvent for PDMS. The

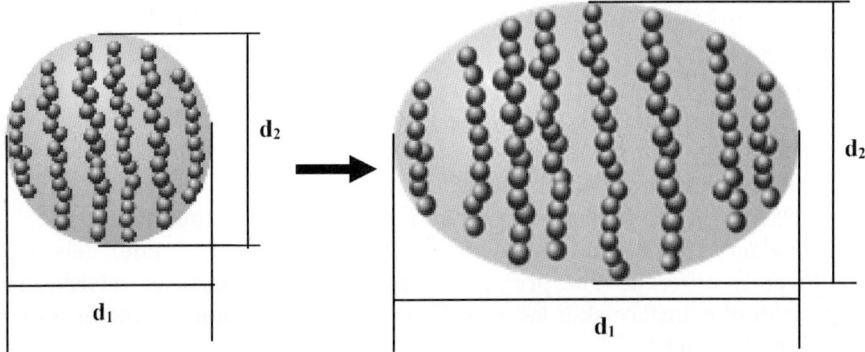

Fig. 42 Anisotropy of swelling is characterized by two linear sizes of the swelling beads perpendicular and parallel to the direction of particle chains. The *arrows* represent the direction of the uniform magnetic field during preparation

Fig. 43 Anisotropic swelling of magnetite-loaded mPDMS in cyclohexane. The *arrows* indicate the direction of the magnetic field (particle chains) during the preparation. The *numbers* in figure represent the lengths that were used to calculate the anisometric quotient

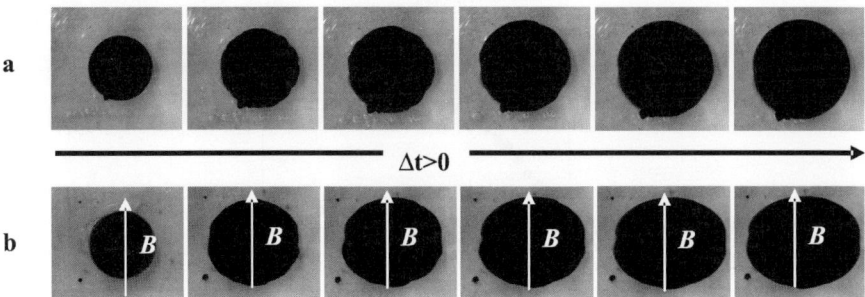

Fig. 44 Swelling kinetics of **a** iron- and **b** magnetite-loaded mPDMS spheres. The *arrow* shows the direction of the applied magnetic field during preparation

swelling degree increased and it was found that the swelling degree parallel to the chain-like particle orientation was less than in the perpendicular case.

In order to characterize the measure of swelling anisotropy, the anisometric quotient (AQ) has been introduced:

$$AQ = \left[1/2(a' + c')/b'[1/2(a + c)/b]\right], \tag{27}$$

where a, b, c are the side lengths of the cube-shaped sample before swelling, and a', b' and c' represent the corresponding lengths after equilibrium has been attained. We have found that the anisometric quotient depends on the magnetite content as well as on cross-linking density. It varies from 1.0 (isotropic sample) to 1.4 (anisotropic gel).

Similar anisotropic volume change has been observed for swelling gel spheres, as shown in Fig. 44. It is rather a surprising finding that magnetite as well as iron-loaded mPDMS samples have strong mechanical anisotropy, but do not show similar swelling characteristics. The different swelling behavior can be explained by the diverse polymer–particle interactions and the surface properties of the filler particles. The iron-loaded samples swells isotropically, as shown in Fig. 44a.

This emphasizes the importance of surface effect between the polymer chains and the surface of solid particles. Since magnetite-loaded mPDMS gels swell anisotropically, as shown in Fig. 44b, one can conclude that the interaction between PDMS chains and the oxide (magnetite) surface is much stronger than the interaction of polymer chains with a smooth metal surface.

8.3
Swelling Kinetics of Unloaded and Magnetite-Loaded PDMS Networks

For the sake of comparison, we have measured the swelling kinetics of the unloaded and isotropic mPDMS networks in cyclohexane. Cyclohexane is a good solvent for PDMS. The diameter of the gel beads was monitored over time. Figure 45 shows the time evolution of the bead size during swelling.

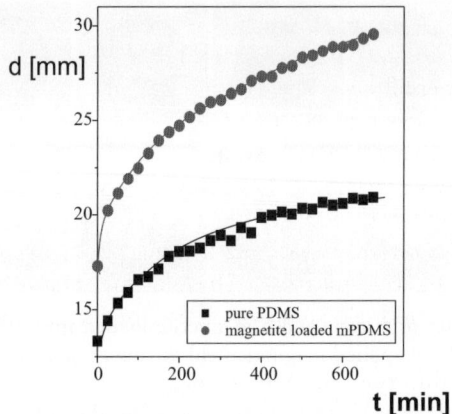

Fig. 45 Swelling kinetics of the unloaded and magnetite-loaded isotropic PDMS networks in cyclohexane. The cross-linker content of the samples was 3.5 wt % and the magnetite content of the isotropic **mPDMS** was 30 wt %, respectively

It can be seen on the figure that there is around a twofold increase in the diameter, which corresponds to eightfold swelling.

We have also measured the swelling kinetics of anisotropic samples, loaded with either magnetite or iron particles. Figure 46 shows the swelling kinetics of magnetite-loaded **mPDMS** beads. The swelling kinetics of iron-loaded samples is also presented.

The time evolution of two perpendicular diameters (d_1 and d_2) were measured. Both the magnetite-loaded and iron-loaded samples show strong me-

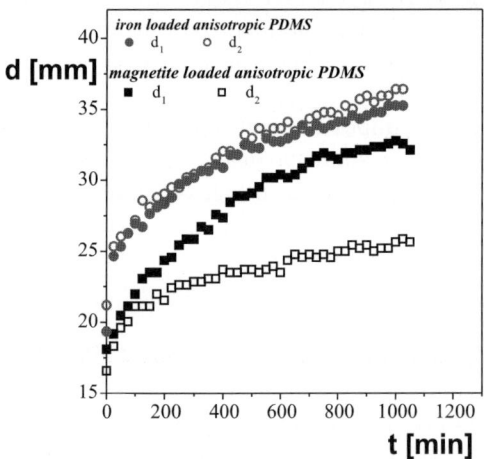

Fig. 46 Swelling kinetics of magnetite- and iron-loaded anisotropic PDMS beads in cyclohexane. The filler content is 30 wt %, and the cross-linker content is 3.5 wt %, respectively

chanical anisotropy. Despite this fact, the swelling behavior is significantly different. The iron-loaded mPDMS samples show no anisotropic swelling. Within the experimental accuracy, the parallel and perpendicular diameters show the same time dependence (Fig. 46). This is not the case for magnetite-loaded samples. It can be seen that the time dependencies of d_1 and d_2 dimensions are very different. At the end of swelling, the diameter d_2 (which is parallel to the particle formation of the egg-shaped sample) is smaller by 35% than d_1. This is due to the strong mutual particle–particle interactions in the pearl chains of solid particles. These attractive interactions retard the swelling in this direction. No such restriction occurs in perpendicular directions.

Comparing the different swelling kinetics of iron- and magnetite-loaded mPDMS networks, it is evident that not only the filler–filler interactions have a strong influence on the swelling behavior, but also the surface properties. The observed difference in swelling kinetics emphasizes the importance of interface effects between the PDMS chains and the surface of the solid particles. The interaction between PDMS chains and rough oxide surface is much stronger than the interaction between a smooth iron surface and polymer chains.

9
Summary of Main Results

The main purpose of the present work was to perform systematic investigations on magnetic field-responsive polymer composites. The ability of magnetic field-sensitive composites to undergo a quick controllable change of shape has been demonstrated. The peculiar magneto-elastic properties may be used to create a wide range of motions, and to control the shape change and movements, which are smooth and gentle similar to those observed in muscle. Thus, application of magnetic field-sensitive composites as a soft actuator for robots and other devices has special interest.

We have prepared magnetoelasts with randomly distributed carbonyl iron particles. It was found that the elastic modulus of magnetoelasts can be increased by an external magnetic field. This effect is called *temporary reinforcement*. A slight increase of elastic modulus by increasing the magnetic field was observed. In order to enhance the magnetic reinforcement effect, we prepared anisotropic samples by varying the spatial distribution of the magnetic particles in the elastic matrix. It was shown that uniaxial field-structured composites exhibit much larger increase in modulus than random particle dispersions. Mechanical properties like the elastic modulus, and the stress–strain behavior of samples characterized by parallel and perpendicular chain-like structures are significantly different. It was found that the temporary reinforcement effect was most significant if the applied field, the particle alignment, and the mechanical stress are all parallel to each other. A phenomenological approach was proposed to describe the dependence of the elastic modulus on the magnetic induction. These magnetic field-sensitive materials with tuneable elastic properties may find usage in elastomer bearings and vibration absorbers.

We have studied the effect of magnetic nanoparticles on the collapse transition of chemically cross-linked NIPA gels. Both the concentration of the magnetic particles, as well as the cross-linking density was varied. It was found that the incorporated magnetite particles decrease the equilibrium swelling degree but do not shift the collapse transition temperature. Below the phase transition temperature, the mass swelling degree increased with decreasing the cross-linking ratio. Above the phase transition temperature, the cross-linking density does not alter the swelling degree. Within the experimental accuracy, the relative swelling degree was found to be a unique function of the temperature. In contrast to this behavior, the temperature dependence of the mass swelling degree was influenced by the cross-linking degree of NIPA gels. Shrinking kinetics measurements were also performed in order to study the influence of the surface skin layer. The kinetics of shrinking of poly(N-isopropylacryalmide) and magnetic poly(N-isopropylacryalmide) gels are significantly different. The presence of magnetic nanoparticles retards the formation of a surface skin layer and bubble formation and results in faster shrinking kinetics.

We have prepared PDMS composite gels with different cross-linking densities. Magnetite and iron particles were built into the network either in randomly distributed form or in a chain-like structure. The composite gels show anisotropic behavior in terms of both mechanical and swelling properties. While the consideration that the shear and longitudinal modulus affect the diffusion is fully legitimate for unloaded polymer gels, it is not so for composite gels where the matrix adheres to the filler. These constraints are responsible for generation of stress and density (and, therefore, degrees of swelling and moduli) fields which may also affect the diffusion. Despite the mechanical anisotropy observed for both magnetite-loaded and iron-loaded samples, only the magnetite-loaded PDMS gel beads shows anisotropic swelling, characterized by direction-dependent relaxation time.

Acknowledgements This research was supported by the Intel KKK (GVOP-3.2.2-2004-07-0006/3.0), NKFP-3A/081/04, and the Hungarian National Research Fund (OTKA, Grant No. T038228 and F046461). This research is sponsored by NATO's Scientific Division in the framework of the Science for Peace Programme (NATO SFP 977998). The authors render special tanks to Dr. Elena Kramarenko, Prof. Lev V. Nikitin and Dr. Gennadiy Stepanov for the fruitful discussion on the subject of the magnetic field controlled elastic properties. We would like to thank all those who assisted to make our former research project a success: Dr. László Barsi, Dr. Dénes Szabó and Dr. Zsolt Varga.

References

1. Gandhi MV, Thompson BS (1992) Smart materials and structures. Chapman & Hall, UK
2. Okano T (ed) (1998) Biorelated polymers and gels. Academic, New York
3. Dusek K, Patterson D (1968) J Polym Sci A-2 6:1209
4. Dusek K, Prins W (1969) Adv Polym Sci 6:1
5. Hoffman AS (1995) Macromol Symp 98:645
6. Osada Y, Ross-Murphy SB (1993) Scientific American, May 1993, pp 82–87
7. Rossi DE, Kawana K, Osada Y, Yamauchi A (eds) (1991) Polymer gels, fundamentals and biomedical applications. Plenum, New York
8. Verdaguer M (1996) Science 272:698
9. Sato O, Iyoda T, Fujishima A, Hashimoto K (1996) Science 272:704
10. Miller JS, Eptein AJ (1998) Chem Commun, p 1319
11. Carlson JD, Jolly MR (2000) Mechatronics 10:555
12. Ginder JM, Davis LC (1994) Appl Phys Lett 65(26):3410
13. Ginder JM, Nichols ME, Elie LD, Tardiff JL (1999) Proc SPIE 3675:131
14. Ginder JM, Clark SM, Schlotter WF, Nichols E (2002) Int J Modern Phys B 16:17: 18:2412
15. Shiga T, Okada A, Karauchi T (1993) Macromolecules 26:6958
16. Shiga T, Okada A, Karauchi T (1995) J Appl Polym Sci 58:787
17. Jolly MR, Carlson JD, Munoz BC, Bullions TA (1996) J Int Mater Syst Struct 7:613
18. Bossis G, Coquelle E, Kuzhir P (2004) Ann Chim Sci Mat 29:43
19. Bellan C, Bossis G (2002) Int J Modern Phys B 16:2447
20. Hu Y, Wang YL, Gong XL, Gong XQ, Zhang XZ, Jiang WQ, Zhang PQ, Chen ZY (2005) Polym Test 24:324
21. Kaleta J, Lewandowski D, Zajac P (2005) Mater Sci Forum 482:403
22. Zhou GY, Jiang ZY (2004) Smart Mater Struct 13:309
23. Bernadek S (1997) J Magnet Magnet Mater 166:91
24. Bernadek S (1999) Appl Phys A 68:63
25. Dorfmann A, Ogden RW (2004) Q JI Mech Appl Math 57(4):599
26. Farshad M, Benine A (2004) Polym Test 23:343
27. Barsi L, Büki A, Szabó D, Zrínyi M (1996) Progr Colloid Polym Sci, p 102
28. Zrínyi M, Barsi L, Büki A (1996) J Chem Phys 104(20):8750
29. Zrínyi M, Barsi L, Büki A (1997) Polym Gels Networks 5:415
30. Szabó D, Barsi L, Büki A, Zrínyi M (1997) Models Chem 134(2):155
31. Zrinyi M (1997) Trends Polym Sci 5:280
32. Zrínyi M, Barsi L, Szabó D, Kilian HG (1997) J Chem Phys 108(13):5685
33. Szabó D, Szeghy G, Zrinyi M (1998) Macromolecules 31:6541
34. Barsi L, Zrinyi M (1998) ACH-Models Chem 153(3):241
35. Zrinyi M, Szabo D, Barsi L (1998) J Intell Mater Sys Struct 9:667
36. Zrinyi M, Szabo D, Kilian HG (1999) Polym Gel Networks 6:6:441
37. Mitsumata T, Ikeda K, Gong JP, Osada Y, Szabó D, Zrínyi M (1999) J Appl Phys 85:12:1
38. Zrínyi M (2000) Colloid Polym Sci 27:2:98
39. Szabo D, Czako-Nagy I, Zrinyi M, Vertes A (2000) J Colloid Interface Sci 221:166
40. Török Gy, Lebedev VT, Cser L, Zrínyi M (2000) Physica B 396:276
41. Zrínyi M, Szabó D, Barsi L (1999) In: Osada Y, Rossi DE (eds) Magnetic field sensitive polymeric actuators. Polymer sensors and actuators. Springer, Berlin Heidelberg New York, p 385

42. Zrínyi M, Szabó D, Filipcsei G, Fehér J (2002) In: Osada Y, Khokhlov A, Dekker M (eds) Electric and magnetic field sensitive smart polymer gels. Polymer gels and networks. CHIPS, New York, p 309
43. Raikher YL, Stolbov OV (2003) J Magnet Magnet Mater 477:258
44. Farshad M, Roux ML (2005) Polym Test 24:163
45. Starodubtsev SG, Saenko EV, Dokukin ME, Aksenov VL, Klechkovskaya VV, Zanaveskina IS, Khokhlov AR (2005) J Phys: Condens Matter 17:1471
46. Hernández R, Sarafian A, López D, Mijangos C (2004) Polymer 46:5543
47. Raikher YL, Stolbov OV (2005) J Magnet Magnet Mater 289:62
48. Jarkova E, Vilgis TA (2004) Macromol Theory Simul 13:592
49. Mayer CR, Cabuil V, Lalot T, Thouvenot R (2000) Adv Mater 12(6):417
50. Teixeira AV, Morfin I, Ehrburger-Dolle F, Rochas C, Geissler E, Licinio P, Panine P (2003) Phys Rev E 67:021504
51. Bohlius S, Brand HR, Pleiner H (2004) Phys Rev E 70:061411
52. Raj K, Moskowitz R (1990) J Magnet Magnet Mater 85:233
53. Rosenweig RE (1985) Ferrohydrodynamics. Cambridge University Press, Cambridge
54. Si S, Li C, Wang X, Yu D, Peng Q, Li Y (2005) Crystal Growth Des 5(2):391
55. Berkovsky BM, Bashtovoy V (eds) (1996) Magnetic fluids and applications handbook. Begell House, New York
56. Otaigbe JU, Barnes MD, Fukui K, Sumter BG, Noid DW (2001) Adv Polym Sci 154:1
57. Nakano M, Koyama K (eds) (1997) Electro-rheological fluids, magneto-rheological suspensions and their applications. World Scientific, Hackensack, NJ
58. Panhurst QA, Connolly J, Jones SK, Dobson J (2003) J Phys D Appl Phys 36:167
59. Wormuth K (2001) J Coll Inter Sci 241:366
60. Ramírez LP, Landfester K (2003) Macromol Chem Phys 204:22
61. Nishio Y, Yamada A, Ezaki K, Miyashita Y, Furukawa H, Horie K (2004) Polymer 45:7129
62. Chatterjee J, Haik Y, Chen CJ (2001) Colloid Polym Sci 279:1073
63. Ma Z, Guan Y, Liu H (2005) J Polym Sci A: Polym Chem 43:3433
64. Iacob GH, Rotariu O, Strachan NJC, Hafeli UO (2004) Biorheology 41:599
65. Zhou SQ, Chu B (1998) J Phys Chem B 102:1364
66. Jones CD, Lyon LA (2000) Macromolecules 33:8301
67. Chatterjee J, Haik Y, Chen CJ (2003) J Appl Polym Sci 91:3337
68. Deng Y, Yang W, Wang C, Fu S (2003) Adv Mater 15:1729
69. Sauzedde F, Elaissari A, Pichot C (1999) Colloid Polym Sci 277:846
70. Xulu M, Filipcsei G, Zrínyi M (2000) Macromolecules 33(5):1716
71. Gilányi T, Varga I, Mészáros R, Filipcsei G, Zrínyi M (2001) Langmuir 17(16):4764
72. Gilányi T, Varga I, Mészáros R, Filipcsei G, Zrínyi M (2001) J Phys Chem B 105(38):971
73. Kondo A, Fukuda H (1999) Colloids Surf A: Physicochem Eng Aspects 153:435
74. Mark JE (1985) British Polym J 17:144
75. Haas W, Zrínyi M, Kilian HG, Heise B (1993) Colloid Polym Sci 271:1024
76. Filipcsei G, Szilágyi A, Csetneki I, Zrínyi M (2006) Polym Adv Technol 239:130
77. Varga Z, Filipcsei G, Szilágyi A, Zrínyi M (2005) Macromol Symp 227:123
78. Varga Z, Filipcsei G, Zrínyi M (2006) Polymer 47(1):227
79. Landfester K, Willert M, Antonietti M (2000) Macromolecules 33:2370
80. Csetneki I, Kabai Faix M, Szilágyi A, Kovács AL, Németh Z, Zrínyi M (2004) J Polym Sci A: Polym Chem 42:482
81. Park TG, Choi HK (1998) Macromol Rapid Commun 19:167
82. Csetneki I, Filipcsei G, Zrínyi M (2005) Macromolecules 39:1942

83. Neél L (1949) Geophys A 5:99
84. Mark JE, Erman B (1988) Rubberlike elasticity, a molecular primer. Wiley, NY
85. Flory PJ (1953) Principles of polymer chemistry. Cornell University Press, NY
86. Treloar LRG (1949) The physics of rubber elasticity. Oxford, Clarendon Press
87. Rothon R (1995) Particulate-filled polymer composites. Longman Sci Techn, Harlow, UK
88. Odgen RW (1984) Non-linear elastic deformations. Ellis Horwood, Chichester
89. Tanaka T, Fillmore D (1979) J Chem Phys 70:1214
90. Tanaka T, Nishio I, Sun ST, Uenonishio S (1982) Science 218:467
91. Hirotsu S (1993) Adv Polym Sci 110:1
92. Tanaka T, Fillmore D (1979) J Chem Phys 70:1214
93. Peters A, Candau SJ (1986) Macromolecules 19:1952
94. Peters A, Candau SJ (1988) Macromolecules 21:2278
95. Onuki A (1988) Phys Rev A 38:2192
96. Li Y, Tanaka T (1990) J Chem Phys 92:1365
97. Varga Z, Filipcsei G, Zrínyi M (2005) Polymer 46:7779
98. Nikitin L, Stepanov G, Mironova L, Samus A (2003) J Mag Mag Mater 258–259:468
99. Nikitin L, Mironova L, Kornev K, Stepanov G (2004) Polym Sci A 46(3):301
100. Nikitin L, Stepanov G, Mironova L, Gorbunov A (2004) J Mag Mag Mater 272–276: 2072
101. Abramchuk S, Grishin D, Kramarenko E, Stepanov G, Khokhlov A (2006) Polym Sci A (in press)

Editor: S. Kobayashi

Molecular Imprinting: A Versatile Tool for Separation, Sensors and Catalysis

Wuke Li · Songjun Li (✉)

Key Laboratory of Pesticide & Chemical Biology of Ministry of Education,
College of Chemistry, Central China Normal University, 430079 Wuhan, P.R. China
Lsjchem@msn.com

1	Introduction	191
2	Intermolecular Interaction and Self-Assembly	193
3	Tailor-Made Separation Materials	195
4	Substrate-Selective Sensors	198
5	Alternative Catalysis Materials	201
6	Final Remarks	205
	References	206

Abstract Molecular imprinting is a promising technique for the preparation of polymers with predetermined selectivity and high affinity. Normally, based on the self-assembly of functional monomers and templates (i.e., imprint molecules), the imprinted polymers are produced by crosslinking polymerizations. The templates are subsequently removed from the polymer, leaving behind binding sites complementary to the imprint species in terms of the shape and the position of functional groups. Recognition of the polymer constitutes an induced molecular memory, which makes the binding sites capable of selectively recognizing the imprint species. This article presents a limited review on molecular self-assembly and the uses of these imprinted polymers in separation, sensors, and catalysis. Other aspects including related backgrounds are also discussed.

Keywords Catalysis · Molecular imprinting · Self-assembly · Sensor · Separation

1 Introduction

The concept of developing molecularly imprinted polymers (MIPs) with the capability of recognizing desired molecules selectively has been discussed for quite some time. However, it is only in recent years that this concept has become realized [1, 2]. For this, the most important contributions can be attributed to Dickey [3], Wulff [4], Mosbach [5] and their groups. Inspired by the formation mechanism of antibody as proposed by Pauling [6],

Dickey found that a selective adsorbent for methyl orange could be prepared when using this molecule as guidable template in the synthesis of silica gel (Fig. 1). As he observed, the rebinding showed more than twice as much methyl orange as ethyl orange. However, when ethyl orange was used as the template in the synthesis, the rebinding showed a preferential adsorption for ethyl orange. Based on this observation, Dickey perceived that an imprinted cavity with the shape complementary to the template was formed in the process of polymerization. Subsequently, Wulff and Mosbach's groups advanced this field significantly by showing that efficient molecular recognition could be achieved by introducing functional groups to these binding sites. In the work of Wulff et al. [4], the reversible covalent approach was developed. The monosaccharide template phenyl-α-D-mannopyranoside was first esterified with 4-vinyl phenyl boronic acid, followed by copolymerization with styrene/divinylbenzene (DVB) or ethylene glycol dimethacrylate (EDMA) to form a highly crosslinked porous network MIP. The template was then removed by an aqueous wash, resulting in the binding sites, which exhibited high affinity and selectivity for the template enantiomer. The noncovalent approach, probably the most widely applied approach at present, was pioneered by Mosbach and his group [5, 7–9]. Based on free-radical polymerization, this approach uses the non-covalent self-assembly between template and functional monomers. Recently, some attempts were also made in the combination of the covalent and non-covalent imprints or semi-covalent ap-

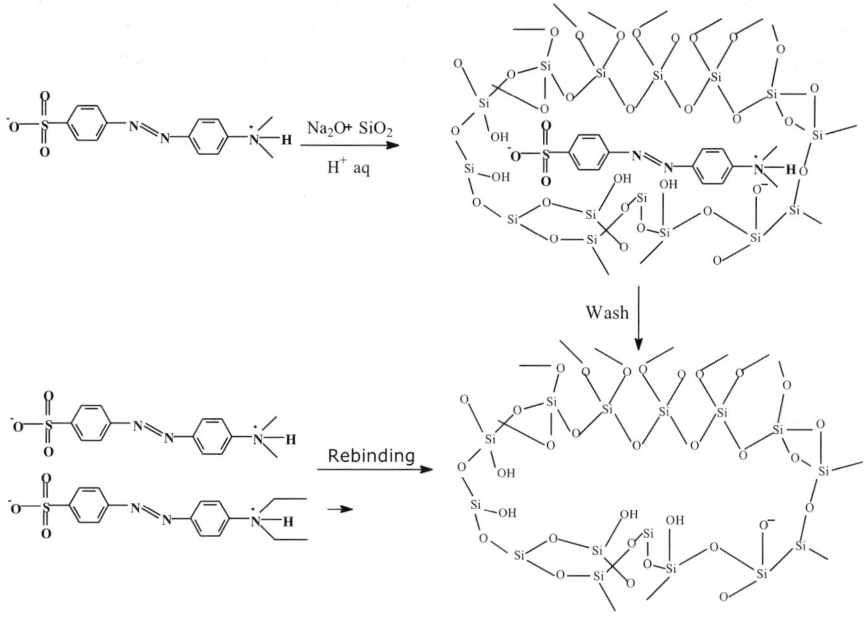

Fig. 1 Selective adsorbent for methyl orange developed by Dickey

proaches [10–13]. Depending on the interactions involved by the imprinting and rebinding, these methods function with either alternate or simultaneous use of covalent and non-covalent interactions.

Currently, the preparation of MIPs often involves three steps:

1. The imprint is achieved by arranging polymerizable functional monomers around a guidable template, and the complexes are formed through covalent, non-covalent, or semi-covalent interactions between the template and monomers
2. The complexes are assembled in the liquid phase and fixed by a crosslinking polymerization
3. Removal of the template leaves behind vacant recognition sites that exhibit high affinity for the imprint species

As a highly crosslinked polymer, the obvious advantages offered by MIPs include physical robustness, high strength, and resistance to elevated temperatures and pressures. Also, the imprinted polymers are generally inert to acids, bases, metal ions and organic solvents. For these reasons, MIPs have been developed for a broad range of applications over the past decade. In retrospect, three main types of use can be summarized: (1) as tailor-made separation materials, (2) as sensors in biosensor-like configurations where the polymers are used as a substitute for the corresponding biological materials, and (3) as catalytically active polymers or enzyme mimics in enzyme technology and organic synthesis. Honestly, a book with many chapters is probably not enough to cover details of all applications involving MIPs, due to the sharply increasing reports over the past decade. The present article places an emphasis on discussing the main points. Other aspects, including related backgrounds, are also discussed.

2
Intermolecular Interaction and Self-Assembly

As commonly known, the monomer–template interaction is the premise for preparing MIPs. Normally, depending on this complementary interaction, the imprint is achieved by arranging polymerizable functional monomers around a guidable template. It has been known that self-assembly plays an important role in predetermining the selectivity [14–16]. Too high a monomer–template ratio will render the imprinted polymer with a spacial and steric mismatching, due to the over-abundance of functional groups which are distributed randomly throughout the polymer. Similarly, too low a ratio will yield a polymer with an insufficient quantity of functional groups to achieve a complete self-assembly, which also results in a low selectivity. Clearly, only a proper monomer–template ratio will produce a polymer with the demonstrated selectivity and a relatively low non-specific binding. For evaluation of the stoichiometric point, excellent works have been presented by Svenson and Andersson [16, 17]. Under the monitoring of UV spectroscopy, a series of functional monomers such as methacrylic acid and its analogs were titrated into a dipeptide (template)-chloroform solution (Fig. 2). A titration of monomer results in a shift of the UV absorption band. The shift becomes evident by the increase in the titration. The shift in the absorption band achieves a maximum when the titrated monomer reaches a critical value. Beyond this critical point, no additional shift in the absorption band is observed except for an increase of absorbency. Further information showed that the imprinted polymers prepared with the stoichiometric composition showed the highest selectivity.

So far, a large number of MIPs have been reported [18–21]. The interactions employed in these polymers are often polar in nature, such as hydrogen bonding, ion pairing, ion-dipole or metal ion chelation. Therefore, the strength of these interactions is accentuated usually by a relatively non-polar porogenic solvent. At present, alternative solvents mainly include acetonitrile, chloroform, dichloromethane, toluene, and dimethyl sulfoxide. The choice of functional monomers is intended to facilitate specific interactions with the imprint molecules. Methacrylic acid (MAA) is the most widely used one. The reason for this can be related to the greater versatility of non-covalent interactions with respect to the available modes of interaction and the more favorable kinetic properties of the recognition process [22]. In the assembly phase, MMA can act as a hydrogen-bond donor or acceptor, enabling ionic interactions to amines and hydrogen bonds to amides, carbamates, or orcarboxyls. The crosslinkers used commonly are ethylene glycol dimethacrylate (EDMA) and trimethylolpropane trimethacrylate (TRIM). In the preparation of MIPs, MMA along with template is assembled and polymerized in the presence of a crosslinker. Subsequently, the template is removed by wash-

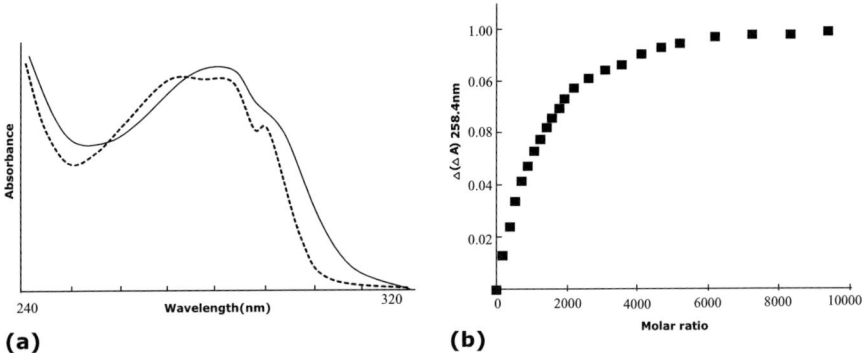

Fig. 2 a Ultraviolet spectra of N-acetyl-L-phenylalaninyl-L-tryptophanyl methyl ester (APTME) in chloroform (*dashed line*) and upon saturation with the monomer analog acetic acid (AA). b Saturation isotherm of titrating APTM with AA

ing from the polymer, leaving behind binding sites complementary to the imprint species in terms of the shape and position of functional groups. The recognition of the polymer constitutes an induced molecular memory, which makes the binding sites capable of selectively recognizing the imprint species.

3
Tailor-Made Separation Materials

As mentioned above, removing imprint molecules from the polymers leaves behind binding sites complementary to the imprint species. Normally, these structures will exhibit a high affinity for the imprint molecules, thereby being quite suitable for separation materials particularly complex separations. The first endeavors based on this approach for enantiomer separations were made by Sellergren et al. [23–26]. With derivatives of amino acid enantiomer as the template, a series of highly selective chiral stationary phases (CSPs) were prepared (Fig. 3). Originally, the template (L-phenylalanine anilide), MMA, and EDMA are dissolved into a low polar medium. The polymerization is then carried out in a glass ampoule using free radical polymerizations, followed by grinding and extracting the template. The resultant particles were sieved to

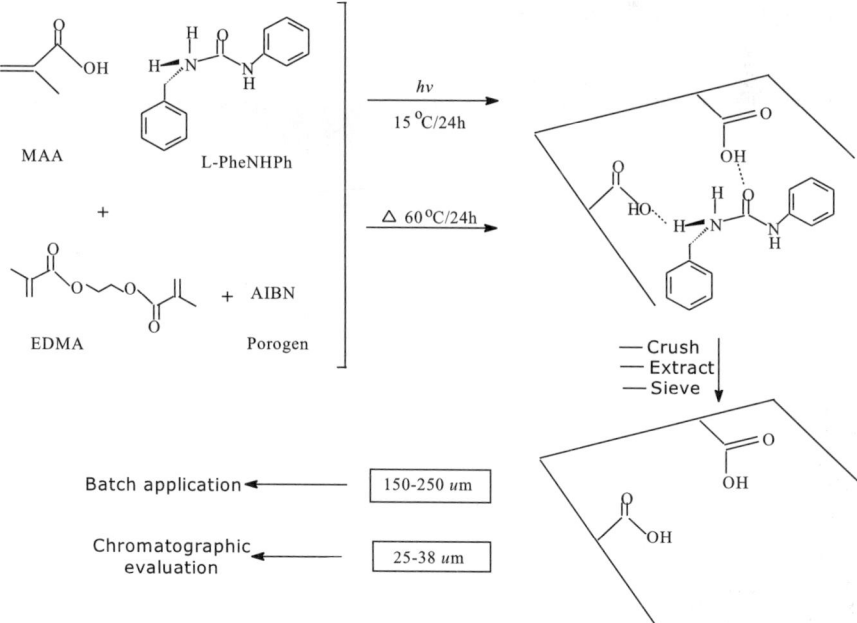

Fig. 3 Schematic presentation of the preparation of Sellergren's MIP

a suitable size for chromatographic (25–38 μm) and batch (150–250 μm) applications. The results showed that the prepared polymers can present a high distinguishability to the enantiomers.

Related works [27–30] further revealed that, when using the basic templates and procedures, an increasing selectivity and affinity could be achieved if increasing the number of proton-accepting or donating sites of the template, or improving the basicity of these sites and the acidity of the functional monomers. In addition, changes in the polymerization conditions so as to lower temperature and the polarity of solvent or increase the pressure and monomer concentration have practical effects on the selectivity [31–33]. These are generally believed to be closely associated with the stability of assembled complexes. In the process of imprinting–polymerization, the change of interaction mode, strength, and polymerization condition can show significant effects on the stability of the assembled complexes and which, in turn, influences the recognition process.

In recent years, significant efforts have gone into the most difficult separations by molecular imprinting. Using N-Ac-L-Phe-L-Trp-OMe as the template and MAA as the functional monomer, Ramstrom et al. [34] have reported that the prepared polymer could easily distinguish this (L,L)-dipeptide from (D,D)-, (D,L)- and (L,D)-dipeptides (α = 47.8, 14.2, and 5.21, respectively). On a similar occasion, Mayes et al. [35] studied the anomeric and epimeric selectivities of MIPs for some glycosides (Fig. 4). The result indicated that the polymers prepared with p-nitrophenyl- or octyl-glycosides could present good anomeric and epimeric selectivities to both of the imprint species. The

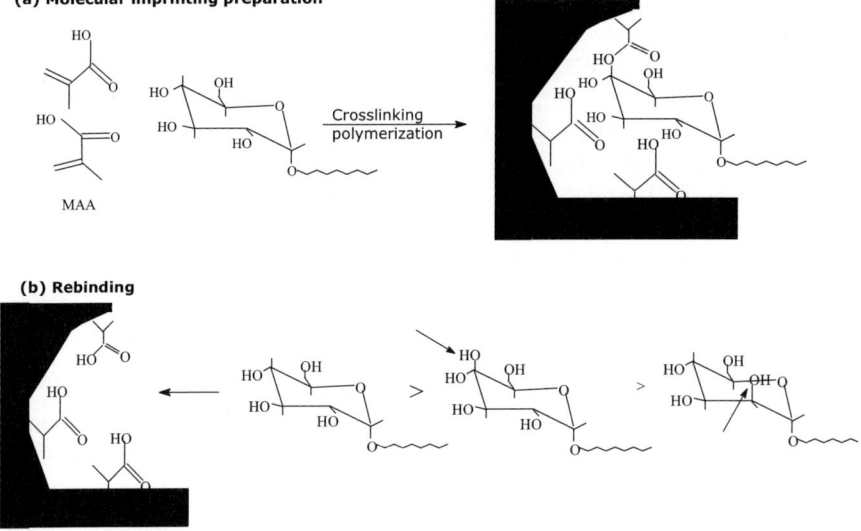

Fig. 4 Preparation (**a**) and selective rebinding (**b**) of Mayes' MIP

insolubility of templates in the non-polar solvent made it difficult to prepare the imprinted polymers with some of these glycosides, particularly methyl-α-D-glucoside. This was solved by imprinting the analog octyl-α-D-glucoside. Further information revealed that the effect of the octyl tail on the imprinting was actually limited. The imprinted polymer could easily distinguish methyl-α-D-glucoside from methyl-α-D-mannoside, methyl-α-D-galactoside, and its p-anomer.

Because of a lack of alternative low-cost approaches, recent works from Li [36] and Donato [37] et al. indicated that molecular imprinting was probably the most straightforward approach for the separation of racemic naproxen. With 4-vinylpyridine as the functional monomer, the imprinted polymer showed a good performance for the chiral separation. Relying on the amount of racemate loaded on the column, separation factors (α) for S- over R-enantiomer, ranging from 1.26 to 1.65, were achieved. The reason why the polymer prepared with 4-vinylpyridine as functional monomer instead of MAA showed better performance may be due to ionic interaction. The ionic interaction is stronger than hydrogen bonding interaction. The introduction of 4-vinylpyridine to the imprint makes the ionic interaction possible. This leads to the longer retention time and thus results in better separation performance. Similarly, the use of 2-vinylpyridine has also been proven efficient for this purpose [38].

Ligand-exchange separation, as noted, is another recent application for molecular imprinting [39, 40]. With the complex [bis-imidazole template-Cu^{2+}-imino diacetic acid monomer] as the imprint, the polymerizations form polymeric coatings. The strong Cu^{2+}-imidazole interaction, useful for high-fidelity imprinting, is too strong for the chromatographic separations of substrate-containing imidazole. However, the metal-complexed polymer can be optimized by replacing Cu^{2+} with another metal ion (e.g., Zn^{2+}) that exhibits faster exchange kinetics and much weaker affinity for imidazole.

As highly selective materials, MIPs have solved many difficult separation problems over the past decade, such as dyes [41], diamines [42], vitamins [43], adrenergic blockers [44], theophylline [45], diazepam [46], nucleotide bases [47], and so on. However, there are also limitations present in conventional MIPs. The separation curves are usually characterized by severe tailing. These are believed to arise from a combination of factors: slow interaction kinetics and the heterogeneous nature of the binding sites with respect to geometry and accessibility [48]. In practice, the tailing can be partially suppressed by conducting separations at an elevated temperature, or by elution. This problem is exacerbated in the majority of early works on developing MIPs, which relied upon the bulk polymerization approach (which usually yielded irregularly shaped particles). As a result, low plate numbers were usually achieved in these MIP-based separations. To overcome the disadvantage, some methods have been proposed for developing uniform MIP particles. These include dispersion polymerization [49], suspension polymerization [50], precipitation polymerization [51], surface imprinting [52], and seed polymerization [53]. Meanwhile, some alternative means have also been employed, in which MIPs are prepared in situ as continuous rods and monoliths [54]. The advantage of in situ methods is that the column packing, a time-consuming and difficult process, is not necessary. In capillary electrochromatography (CEC), the use of MIP-based stationary phases is especially attractive owing to the improved efficiency with respect to separation application.

4
Substrate-Selective Sensors

With regard to substrate-selective sensors with pre-organized cavities, impressive advances have been made in molecular imprinting [55–57]. The discovery of MIP-membrane electroconductivity was an interesting issue, which actually led to the appearance of the earliest MIP sensors [58, 59]. It was shown that the membrane electroconductivity could be a function of the interaction between MIP-membrane and ligand (i.e., imprint species) (Fig. 5). An increase in the ligand concentration would result in an enhancement of membrane conductivity. With the same level of concentration, a maximal electroconductivity with the imprint species could be achieved. In addition, it has also been confirmed that polymers imprinted with amino acids, nucleosides, atrazines, sialic acids, or cholesterols can show similar features if coupled with the appropriate transducer [60–64]. In particular, molecular imprinting is presently probably the only choice when no suit-

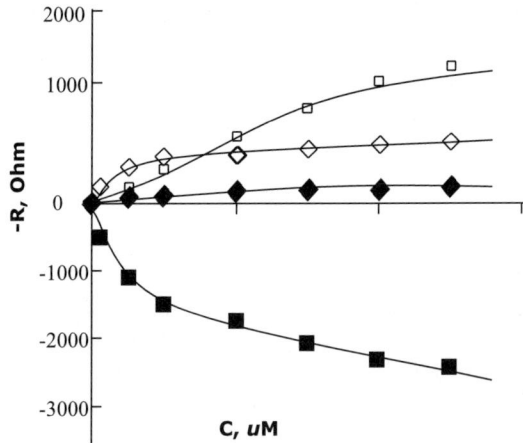

Fig. 5 Dependency of MIP electroconductivity on ligand concentration: ♦ aminopropyl uracile, ◊ atrazine, □ L-Phe, ■ sialic acid

able biomolecule is available, such as in the screening of orphan receptors etc. Compared to to conventional biosensors, the merits offered by MIP sensors include drastically improved thermal, mechanical, and chemical stability. Thus, MIP sensors are attractive for various convenient applications.

Based on the observation that chloramphenicol (CAP)-imprinted polymer possessed a modest affinity for chloramphenicol-methyl red (CAP-MR), Levi et al. [65] designed an intriguing MIP sensor to monitor the change of CAP in patients' blood (Fig. 6). The presence of CAP in blood leads to a competitive displacement of CAP-MR from the imprinted cavities. The displaced composite is subsequently monitored at 460 nm. After optimizing the flow rate and concentration of CAP-MR in acetonitrile mobile phase, the response of this system to CAP, thiamphenicol (TAM), and chloramphenicol diacetate (CAP-DA) was determined (Fig. 7). As observed for CAP, there was a linear correlation over the range 1–1000 µg/mL. However, for CAP-DA almost no appreciable response was achieved, even if it was injected to 1000 µg/mL. As also observed, the value for CAP was about 40% higher than that for TAM at the same concentration. This revealed that CAP could compete more efficiently with the bound CAP-MR than TAM did. Further information showed that this method was adequate for detection below and above the recommended therapeutic range (10–20 µg/mL serum, potentially toxic above 25 µg/mL).

For convenient application, McNiven et al. [66] have simplified the system by introducing an in situ imprinting technique into the system (wherein

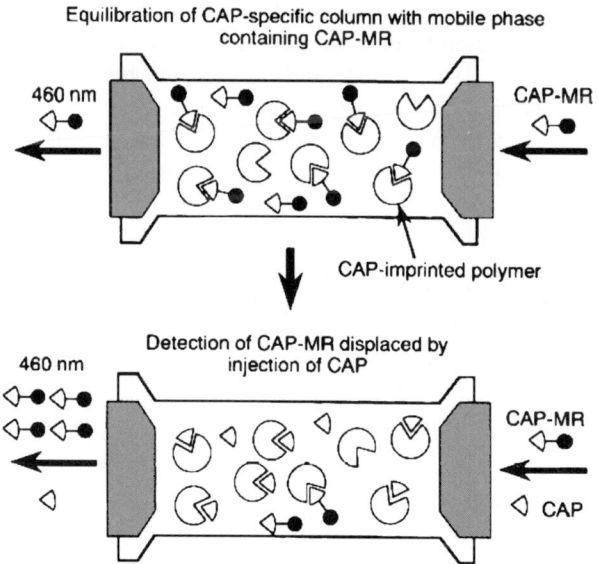

Fig. 6 MIP sensor developed by Levi

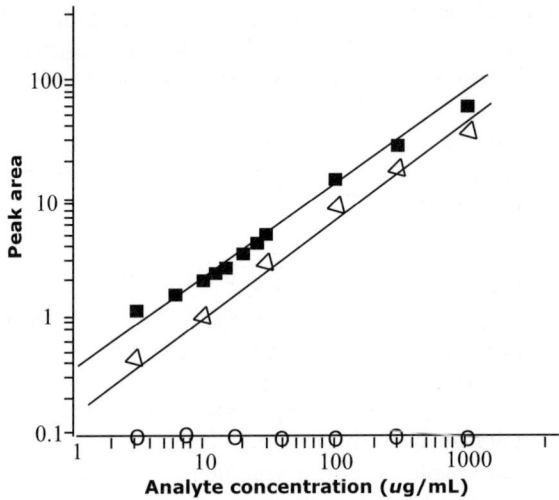

Fig. 7 Response of MIP sensor to CAP (■), TAM (□), and CAP-DA (○)

the polymer is formed inside a chromatography column). Though narrower in the response range (3–30 μg/mL), the column performance was comparable with that of bulk-polymerized materials. Recently, in order to monitor atrazines (a kind of herbicide) in drinking water, foods, and environments, Piletsky et al. [67] designed a MIP sensor. With atrazine as the template, and diethylaminoethyl methacrylate (DEAEM) and MAA as the functional monomer, a series of imprinted membranes were synthesized. A small amplitude-alternating voltage was subsequently used to record the electro-resistances of these membranes in both the presence and absence of the imprinted molecules. As they observed, the electro-resistances of these prepared membranes were closely associated with the atrazine concentration. An increase in the atrazine concentration would result in a decrease of membrane electro-resistances. When using DEAEM as the functional monomer and at a 10 : 3 monomer:template ratio, the highest selectivity for atrazine recognition could be obtained. It was also showed that these prepared membranes could be stored at room temperature for 4 months without any loss of sensitivity. Based on identical technology, MIP sensors showing responses to 6-amino-1-propyluracil and L-phenylalanine were also constituted recently [68].

Over the past decade, there have been some MIP-based sensors available as transducers converting the signals of polymer recognition into physical ones (Table 1) [68]. Also considerable efforts have been made on electrochemical detection [69–71]. However, there is much room for improvement regardless of the expected advantages and great perspectives. As known, there is still substantial difficulty in preparing stable MIP membranes and thin layers with reproducible properties from highly crosslinked polymers. Deterioration due

Table 1 Summary of MIP sensors

Class	Typical analyte	Functional monomer	Detection range
Fluorimetry	Triazine	MAA	0.01–100 mM
	Sialic acid	Allylamine + TVPhB	0.5–10 µM
	Dansyl-L-phenylalanine	MAA, 2VPy	0–30 µg/mL
	Pyrene	Aromatic polyurethane	0–40 µg/L
	cAMP	DMASVBP + HEMA	0.1–100 µM
	NATA	HEAPTES + TES	Qualitative
Conductometriy	Atrazine	DEAEM	0.01–0.5 mg/L
	Sialic acid	Allylamine + TVPhB	1–50 µM
	Morphine	MAA	Qualitative
	L-Phenylalanine	DEAEM	0.05–0.4 mM
	Chloramphenicol	DEAEM	1–1000 µg/mL
	Chloramphenicol	DEAEM	3–30 µg/mL
	Testosterone	MAA	0.1–1.25 mM
Potentiometry	Phenylalanine anilide	MAA	33–3300 µg/mL
Capacitance	Phenylalanine anilide	MAA	Qualitative
Amperometry	Morphine	MAA	0.1–10 µg/mL
SAW, QMB	o-Xylene	Aromatic polyurethane	Qualitative
Luminescence	PMP	Eu + DVMB	0.125–150 000 µg/L
pH	Glucose	STACNCu	0–25 mM
SPR	Theophyline	MAA	0.4–6 mg/mL

to this problem is the reason why most reported works have used relatively excessive crosslinkers, which usually yield an exorbitantly high crosslinkage. Also, in fact, mass transfer in the imprint cavities is usually not sufficient for practical application if these cavities are exclusively or predominately induced by the imprint. Thus, significant modification of polymer preparation had to be made before receiving more commercial interest. Alternative methods for this currently include surface imprinting [72], ultrathin film (UF) MIP composite, or asymmetric UF membrane [73] etc. Regardless of these, practical MIP sensors need still more work.

5
Alternative Catalysis Materials

Recently, some groups have begun to utilize MIPs as active materials for catalyzing some reactions [74–78]. Although MIPs are unlikely to outperform enzymes or even catalytic antibodies in activity and selectivity, they can work in organic solvents and under harsh conditions. This makes them useful supplements for alternative catalysis materials. Unlike the mentioned separation materials, the preparation of MIPs as catalysts requires choice of a defined transition state analog (TSA) as the template. This is because the activation energy, in most cases, must pass through a transition state in going from the substrate towards the product. As commonly known, the transition state itself cannot be used as the template, because of its instability. Therefore, a TSA must be found to generate imprints within the polymer matrix. This will sometimes be extremely similar to the transition state and substrate, or in some cases similar in the region of specific functionality with respect to the group distances. Normally, the more similarities there are between TSA and the transition state, the better the activity and selectivity expected. Otherwise, the activation of substrate would be restricted to a lesser extent. To illustrate this, Beach [79] et al. designed some elaborate experiments (Fig. 8). With benzylmalonic acid and acetic acid as TSAs and N-(2-aminoethyl)-methacrylamide as functional monomer, MIP catalysts possessing dehydrofluorination activity of 4-fluoro-4-(p-nitrophenyl) butan-2-one were prepared. The catalytic determination indicated that the polymer prepared with

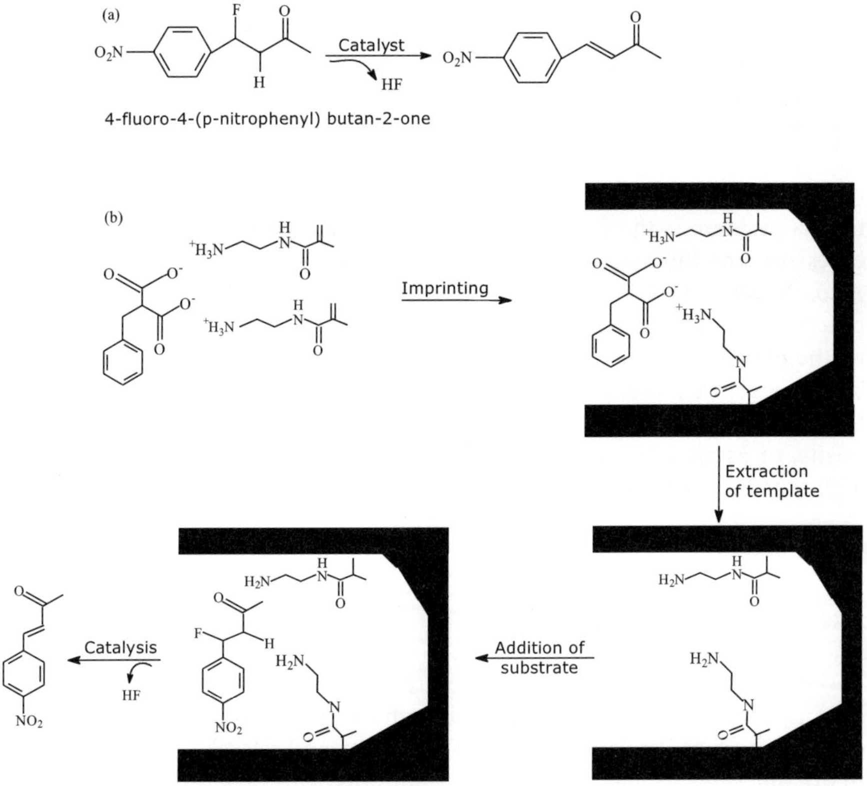

Fig. 8 MIP catalysis system developed by Shea and Mosbach

benzylmalonic acid as the TSA could display a good activity. However, with acetic acid, almost no appreciable activity was achieved. Muller et al. [80] conducted similar investigations. The results also indicated the same information. Related studies further revealed that the shape of imprinted cavities can aid recognition of specific substrates from related structures [81–83]. Thus, it is currently accepted that the use of TSA is necessary for the preparation of MIP catalysts and can play a role in determining the efficiency of designed catalysts.

As noted, some novel strategies were employed recently for developing MIP catalysts for Diels–Alder cycloadditions (Fig. 9) [84, 85]. Interestingly, it was not a specific TSA but the precursor, namely chlorendic anhydride, that was used in these works. In addition, it does not require the use of solvent since the porogenic structure of the resultant polymer can come from the dissolution of the silica support. Indeed, the actual TSA was achieved by covalently coupling the precursor into amino groups on the silica support (Fig. 9). The subsequent process is the acetylation of free amino groups (with acetic anhydride) and the addition of functional monomer (MAA) and crosslinker (DVB), followed by the polymerization. The resultant core-shell composites were treated by hydrofluoric acid to dissolve the silica and remove the template. The catalysis determination revealed that the prepared MIP exhibited a good activity for the mentioned reaction. This was evidenced by the increase of reaction temperature. Meanwhile, a reduction in the activation energy from 63 to 55 kJ mol^{-1} was observed from the prepared MIP cata-

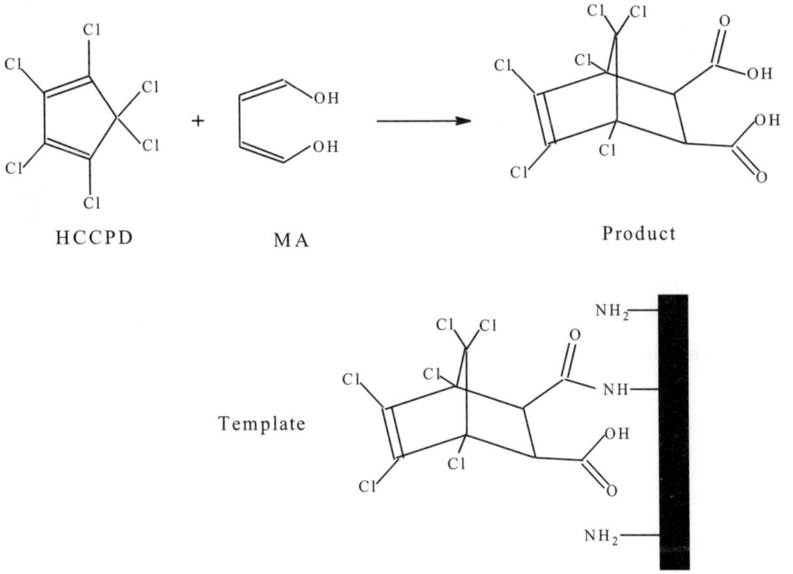

Fig. 9 Preparation of Visnjevski–Yilmaz's MIP catalyst

lyst. Related studies also indicated that the catalyst prepared with the novel strategies was usually more efficient than that prepared by traditional imprinting [86, 87]. The reason for this is generally related to the improvement of binding-site distribution. With these novel strategies, the majority of the binding sites are localized on or near the polymer surface, which increases the accessibility to substrates. Further, it is also anticipated that these strategies can partially aid in improving the performance, particularly the uniformity

Table 2 Important reactions catalyzed by MIPs

TSA	Substrate	Relative catalytic effect of MIP
Hydrolysis		
Pyridin-derivative of N-Boc-aminoacids	Nitrophenyl ester	$K_{imp}/K_{non\text{-}imp} = 4\text{--}5$
p-Nitrophenylmethyl Phosphonate	p-Nitrophenol acetate	$K_{imp}/K_{non\text{-}imp} = 1.6$
Phosphonate	Aminoacid ester	$K_{imp}/K_{non\text{-}imp} = 3$
Phosphonate	Carbonic acid ester	$K_{imp}/K_{non\text{-}imp} \approx 100$
Phosphonate	Aminoacid ester	$K_{imp}/K_{non\text{-}imp} = 2.54$
Dehydrofluorination		
N-benzyl-isopropylamine	4-Fluoro-4-(p-nitrophenyl)-2-butanone	$K_{imp}/K_{non\text{-}imp} < 2.4$
Benzyl-isopropylamine	4-Fluoro-4-(p-nitrophenyl)-2-butanone	$K_{imp}/K_{non\text{-}imp} = 3.2$
N-methyl-N-(4-nitrobenzyl)-δ-aminovaleric acid	4-Fluoro-4-(p-nitrophenyl)-2-butanone	$K_{imp}/K_{non\text{-}imp} = 3.3$
N-(p-Nitrobenzyl)-isopropylamine	4-Fluoro-4-(p-nitrophenyl)-2-butanone	$K_{imp}/K_{non\text{-}imp} = 3.27$
Diels–Alder reaction		
Chlorendic anhydride	Tetrachlorothiophen-dioxide + maleic anhydride	$(K_{imp} - K_{non\text{-}imp})/K_{non\text{-}imp} = 270$
Aldol condensation		
Dibenzoylmethane (DBM) + Co^{2+}	Acetophenone and benzaldehyde	$r_{(DBM/Cu^{2+}\text{-}MIP)}/r_{(Co^{2+}\text{-}MIP)} \leq 2$
Isomerization		
Indol	Benzisoxazol	$K_{imp}/K_{non\text{-}imp} = 7.2$

of attached templates. As also noted, Sode et al. [88, 89] reported a novel polymer catalyst that could work as a fructosylamine dehydrogenase. In the presence of an electron acceptor (i.e., 1-methoxy-5-methylphenazinium methylsulfate), poly(vinylimidazole) functioned as the catalyst for oxidative cleavage of fructosylamine. The reduced electron acceptor was oxidized again over the electrode surface. The produced imine compound was then hydrolyzed into glucosone and valine.

As catalysis materials, significant progress has been made in molecular imprinting over the last few years. Some reported cases include hydrolysis, dehydrofluorination, Diels–Alder reaction, and isomerization (Table 2) [90]. However, the development of practical catalysts is in its infancy. In particular, more work is necessary on the activity and selectivity. Also some inherent problems, such as the preparation of template molecules, their availability as well as the recycling, need still more endeavors. Fortunately, considerable effort is currently being made.

6
Final Remarks

Because of the potential usability, molecular imprinting has attracted much attention over the last decade. Normally, based on the self-assembly of polymerizable monomers and templates, the imprinted polymers (MIPs) are produced by a crosslinking polymerization. The templates are subsequently removed from the polymer, leaving behind binding sites complementary to the imprint species in terms of the shape and the position of functional groups. Recognition of the polymer constitutes an induced molecular memory, which makes the binding sites capable of selectively recognizing the imprint species. This feature makes molecular imprinting quite attractive for various applications. This article presents a limited review on self-assembly and the uses of these imprinted polymers. As commonly known, self-assembly plays an important role in predetermining the formation of efficient imprinting. Only an optimal ratio could produce the polymer with a demonstrated selectivity and a relatively low non-specific binding. Too high or too low a functional monomer–template ratio will render an MIP with high non-specific adsorption.

As separation materials, the obvious advantage offered by MIPs is a relatively straightforward and predetermined selectivity. Based on molecular imprinting, some difficult separations, particularly enantiomer separations, have been solved. Amino acid enantiomers, chiral dipeptide, and racemic naproxen are examples. In particular, molecular imprinting is probably currently the only choice where no suitable biomolecule is available. When coupling with appropriate transducers such as electro-, photo- or magnetochemical transducers, MIPs may be applied as the monitors in several systems. In contrast to MIPs as separation and sensor materials, MIPs as alternative catalysts require a defined transition state analog as the template. The reason for this is that, in most cases, the activation energy must pass through a transition state in going from the substrate towards the product.

It is also necessary to point out that there is still signification room for improvement in molecular imprinting. Some inherent problems, including increasing reproducibility of various applications and seeking the availability of templates and their recycling, need still more work. Ongoing efforts to improve the imprinting fidelity and choose or synthesize suitable templates are expected to be helpful.

Acknowledgements The authors want to thank NSFC for presenting financial support to conduct this work (Granted No. 20603010). Thank also Professor Guangfu Yang for providing constructive suggestions in the revision process, which play an important role on improving this article.

References

1. Meng ZH, Sode K (2005) The molecular reaction vessels for a transesterification process created by molecular imprinting technique. J Mol Recogn 18(3):262–266
2. Takeuchi T, Mukawa T, Shinmori H (2005) Signaling molecularly imprinted polymers: molecular recognition-based sensing materials. Chem Record 5(5):263–275
3. Dickey FH (1949) The preparation of specific adsorbents. Proc Natl Acad Sci 35(5): 227–229
4. Wulff G (1995) Molecular imprinting in cross-linking materials with the aid of molecular templates: a way towards artificial antibodies. Angew Chem Int Ed 34(17): 1812–1832
5. Mosbach K, Ramstrom O (1996) The emerging technique of molecular imprinting and its future impact on biotechnology. Biotechnol 14(2):163–170
6. Pauling L, Campbell D (1942) The manufacture of antibodies in vitro. J Experimtl Med 76(2):211–220
7. Mosbach K (2001) Toward the next generation of molecular imprinting with emphasis on the formation, by direct molding, of compounds with biological activity (biomimetics). Anal Chim Act 435(1):3–8
8. Mosbach K, Yu Y, Andersch J, Ye L (2001) Generation of new enzyme inhibitors using imprinted binding sites: the anti-idiotypic approach, a step toward the next generation of molecular imprinting. J Am Chem Soc 123(49):12420–12421
9. Ye L, Mosbach K (2001) Molecularly imprinted microspheres as antibody binding mimics. React Funct Polym 48(2):149–157
10. El-ghayoury A, Hofmeier H, de Ruiter B, Schubert US (2003) Combining covalent and noncovalent cross-linking: a novel terpolymer for two-step curing applications. Macromolecules 36(11):3955–3959
11. Whitcombe MJ (2006) MIP catalysts – from theory to practice. In: Piletsky S, Turner A (eds) Molecular imprinting of polymers. Landes Bioscience 83–108
12. Shiomi T, Matsui M, Mizukami F, Sakaguchi K (2005) A method for the molecular imprinting of hemoglobin on silica surfaces using silanes. Biomaterials 26(27):5564–5571

13. Klein JU, Whitcombe MJ, Mulholland F, Vulfson EN (1999) Template-mediated synthesis of a polymeric receptor specific to amino acid sequences. Angew Chem Int Ed 38(13):2057–2060
14. Fish WP, Ferreira J, Sheardy RD, Snow NH, O'Brien TP (2005) Rational design of an imprinted polymer: maximizing selectivity by optimizing the monomer-template ratio for a cinchonidine MIP, prior to polymerization, using microcalorimetry. J Liquid Chromatogr. Related Technol 28(1):1–15
15. Svenson J, Ning Z, Fohrman U, Nicholls IA (2005) The role of functional monomer-template complexation on the performance of atrazine molecularly imprinted polymers. Anal Lett 38(1):57–69
16. Svenson J, Andersson HS, Piletsky SA, Nicholls IA (1998) Spectroscopic studies of the molecular imprinting self assembly process. J Mol Recogn 11(1):83–86
17. Andersson HS, Nicholls IA (1997) Spectroscopic evaluation of molecular imprinting polymerization systems. Bioorg Chem 25(3):203–211
18. Li Z, Day M, Ding J, Faid K (2005) Synthesis and characterization of functional methacrylate copolymers and their application in molecular imprinting. Macromolecules 38(7):2620–2625
19. Kim TH, Ki CD, Cho H, Chang T, Chang JY (2005) Facile preparation of core-shell type molecularly imprinted particles: molecular imprinting into aromatic polyimide coated on silica spheres. Macromolecules 38(15):6423–6428
20. Ou JJ, Tang SW, Zou HF (2005) Chiral separation of 1,1'-bi-2-naphthol and its analogue on molecular imprinting monolithic columns by HPLC. J Separat Sci 28(17):2282–2287
21. Turiel E, Martin-Esteban A (2005) Molecular imprinting technology in capillary electrochromatography. J Separat Sci 28(8):719–728
22. Kim H, Guiochon G (2005) Thermodynamic studies on the solvent effects in chromatography on molecularly imprinted polymers. 1. Nature of the organic modifier. Anal Chem 77(6):1708–1717
23. Sellergren B (2001) Imprinted chiral stationary phases in high-performance liquid chromatography. J Chromatogr A 906(1):227–252
24. Sellergren B, Lepistoe M, Mosbach K (1988) Highly enantioselective and substrate-selective polymers obtained by molecular imprinting utilizing noncovalent interactions. NMR and chromatographic studies on the nature of recognition. J Am Chem Soc 110(17):5853–5860
25. Sellergren B (1989) Molecular imprinting by noncovalent interactions: tailor-made chiral stationary phases of high selectivity and sample load capacity. Chirality 1(1):63–68
26. Sellergren B (1997) Noncovalent molecular imprinting: antibody-like molecular recognition in polymeric network materials. Trends Anal Chem 16(6):310–320
27. Huang X, Kong L, Li X, Zheng C, Zou H (2003) Molecular imprinting of nitrophenol and hydroxybenzoic acid isomers: effect of molecular structure and acidity on imprinting. J Mol Recogn 16(6):406–411
28. Allender CJ, Brain KR, Heard CM (1997) Binding cross-reactivity of Boc-phenylalanine enantiomers on molecularly imprinted polymers. Chirality 9(3):233–237
29. Fischer L, Mueller R, Ekberg B, Mosbach K (1991) Direct enantioseparation of beta-adrenergic blockers using a chiral stationary phase prepared by molecular imprinting. J Am Chem Soc 113(24):9358–9360
30. Andersson HS, Koch-Schmidt AC, Ohlson S, Mosbach K (1996) Study of the nature of recognition in molecularly imprinted polymers. J Mol Recogn 9(6):675–682

31. Lin JM, Nakagama T, Uchiyama K, Hobo T (1998) Temperature effect on chiral recognition of some amino acids with molecularly imprinted polymer filled capillary electrochromatography. Biomed Chromatogr 11(5):298–302
32. Li ZW, Yang GL, Wang DX, Zhou SL, Liu ED, Chen Y (2003) Separation of aminoantipyrine and its close analogues by molecular imprinting stationary phase. Chem J Internet 5(6):46
33. Gavioli E, Maier NM, Haupt K, Mosbach K, Lindner W (2005) Analyte templating: enhancing the enantioselectivity of chiral selectors upon incorporation into organic polymer environments. Anal Chem 77(15):5009–5018
34. Ramstrom O, Nicholls IA, Mosbach K (1994) Synthetic peptide receptor mimics: highly stereoselective recognition in nonconvalent molecularly imprinted polymers. Tetrahedron: Assymetry 5(4):649–656
35. Mayes AG, Andersson L, Mosbach K (1994) Sugar binding polymers showing high anomeric and epimeric discrimination by noncovalent molecular imprinting. Anal Biochem 222(2):483–488
36. Li P, Rong F, Xie YB, Hu V, Yuan CW (2004) Study on the binding characteristic of S-naproxen imprinted polymer and the interactions between templates and monomers. J Anal Chem 59(10):939–944
37. Donato L, Figoli A, Drioli E (2005) Novel composite poly (4-vinylpyridine)/polypropylene membranes with recognition properties for (S)-naproxen. J Pharm Biomed Anal 37(5):1003–1008
38. Kempe M, Mosbach K (1994) Direct resolution of naproxen on a non-covalently molecularly imprinted chiral stationary phase. J Chromatogr A 664(2):276–279
39. Arnold FH, Plunkett SD, Vidyasankar S (1994) Template polymerization using metal ion coordination: metal replacement to optimize templating and substrate re-binding. Polym Preprints 35(9):996–997
40. Uezu K, Yoshida M, Goto M, Furusaki S (1999) Molecular recognition using surface template polymerization. Chemtech 29(4):12–18
41. Gong SL, Yu ZJ, Meng LZ, Hu L, He YB (2004) Dye-molecular-imprinted polysiloxanes. II. Preparation, characterization, and recognition behavior. J Appl Polym Sci 93(2):637–643
42. Dobashi A, Nishida S, Kurata K, Hamada M (2002) Chiral separation of enantiomeric 1,2-diamines using molecular imprinting method and selectivity enhancement by addition of achiral primary amines into eluents. Anal Sci 18(1):32–35
43. Chen ZD, Nagaoka T (2000) Recognition of vitamin K1 with a molecularly imprinted self-assembled monolayer film. Bunsekikagaku, 49(7):543–546
44. Bitz GG, Schmid MG (2001) Chiral separation by chromatographic and electromigration techniques. Biopharm Drug Dispos 22(7):291–336
45. Cai LS, Wu CY, Mei SR, Zeng ZR (2004) Molecularly imprinted polymer theophylline retention and molecular recognition properties in capillary electrochromatography. Wuhan Univ J Natl Sci 9(3):359–365
46. Alvarez-Lorenzo C, Concheiro A (2003) Effects of surfactants on gel behavior: design implications for drug delivery systems. Am J Drug Deliv 1(2):77–101
47. Sellergren B, Buchel G (2005) Porous, molecularly imprinted polymer and a process for the preparation thereof. US Patent 6,881,804 (B1), Apr. 19, Pages 11
48. Vallano PT, Remcho VT (2000) Highly selective separations by capillary electrochromatography: molecular imprint polymer sorbents. J Chromatogr A 887(2):125–135
49. Sellergren B (1994) Imprinted dispersion polymers: a new class of easily accessible affinity stationary phases. J Chromatogr A 673(1):133–141

50. Kim K, Kim D (2005) High-performance liquid chromatography separation characteristics of molecular-imprinted poly(methacrylic acid) microparticles prepared by suspension polymerization. J Appl Polym Sci 96(1):200–212
51. Ho KC, Yeh WM, Tung TS, Liao JY (2005) Amperometric detection of morphine based on poly(3,4-ethylene dioxythiophene) immobilized molecularly imprinted polymer particles prepared by precipitation polymerization. Anal Chim Act 542(1):90–96
52. Say R, Erdem M, Ersoz A, Turk H, Denizli A (2005) Biomimetic catalysis of an organophosphate by molecularly surface imprinted polymers. Appl Catal A 286(2):221–225
53. Piscopo L, Prandi C, Coppa M, Sparnacci K, Laus M, Lagana A, Curini R, D'Ascenzo G (2002) Uniformly sized molecularly imprinted polymers (MIPs) for 17β-estradiol. Macromol Chem Phys 203(10):1532–1538
54. Schweitz L, Spegel P, Nilsson S (2001) Approaches to molecular imprinting based selectivity in capillary electrochromatography. Electrophoresis 22(19):4053–4063
55. Dickert FL, Lieberzeit P, Tortschanoff M (2000) Molecular imprints as artificial antibodies – a new generation of chemical sensors. Sensors Actuators B 65(2):186–189
56. Gao SH, Wang W, Wang BH (2001) Building fluorescent sensors for carbohydrates using template-directed polymerizations. Bioorg Chem 29(5):308–320
57. Liang CD, Peng H, Zhou AH, Nie LH, Yao SZ (2000) Molecular imprinting polymer coated BAW bio-mimic sensor for direct determination of epinephrine. Anal Chim Act 415(2):135–141
58. Piletsky SA, Panasyuk TL, Piletskaya EV, El'skaya AV, Levi R, Karube I, Wulff G (1998) Imprinted membranes for sensor technology: opposite behavior of covalently and noncovalently imprinted membranes. Macromolecules 31(7):2137–2140
59. Piletsky SA, Butovich IA, Kukhar VP (1992) Design of molecular sensors on the basis of substrate-selective polymer membranes. Zh Anal Khim 47(9):1681–1684
60. Zhou YX, Yu B, Levon K (2003) Potentiometric sensing of chiral amino acids. Chem Mater 15(14):2774–2779
61. Shoji R, Takeuchi T, Kubo I (2003) Atrazine sensor based on molecularly imprinted polymer-modified gold electrode. Anal Chem 75(18):4882–4886
62. Ye L, Mosbach K (2001) Polymers recognizing biomolecules based on a combination of molecular imprinting and proximity scintillation: a new sensor concept. J Am Chem Soc 123(12):2901–2902
63. Chou LCS, Liu CC (2005) Development of a molecular imprinting thick film electrochemical sensor for cholesterol detection. Sensors Actuators B 110(2):204–208
64. Mosbach K, Haupt K (1998) Some new developments and challenges in non-covalent molecular imprinting technology. J Mol Recogn 11(1):62–68
65. Levi R, McNiven S, Piletsky SA, Rachkov A, Cheong SH, Yano K, Karube I (1997) Optical detection of chloramphenicol using molecularly imprinted polymers. Anal Chem 69(11):2017–2021
66. McNiven S, Kato M, Yano K, Karube I (1998) Chloramphenicol sensor based on an in situ imprinted polymer. Anal Chim Act 365(6):69–74
67. Piletsky SA, Piletskaya EV, Elgersma AV, Yano K, Parhometz YP, El'skaya AV, Karube I (1995) Atrazine sensing by molecularly imprinted membranes. Biosens Bioelectr 10(10):959–964
68. Yano K, Karube I (1999) Molecularly imprinted polymers for biosensor applications. Trends Anal Chem 18(3):1999–2004
69. Marx KA (2003) Quartz crystal microbalance: a useful tool for studying thin polymer films and complex biomolecular systems at the solution-surface interface. Biomacromolecules 4(5):1099–1120

70. Piletsky SA, Turner APF (2002) Electrochemical sensors based on molecularly imprinted polymers. Electroanalysis 14(5):317–323
71. Zeng YN, Zheng N, Osborne PG, Li YZ, Chang WB, Wen MJ (2002) Cyclic voltammetry characterization of metal complex imprinted polymer. J Mol Recogn 15(4):204–208
72. Zhou Y, Yu B, Shiu E, Levon K (2004) Potentiometric sensing of chemical warfare agents: surface imprinted polymer integrated with an indium tin oxide electrode. Anal Chem 76(10):2689–2693
73. Rich RL, Myszka DG (2005) Survey of the year 2004 commercial optical biosensor literature. J Mol Recogn 18(6):431–478
74. Liu JQ, Wulff G (2004) Functional mimicry of the active site of carboxypeptidase A by a molecular imprinting strategy: cooperativity of an amidinium and a copper ion in a transition-state imprinted cavity giving rise to high catalytic activity. J Am Chem Soc 126(24):7452–7453
75. Becker JJ, Gagne MR (2004) Exploiting the synergy between coordination chemistry and molecular imprinting in the quest for new catalysts. Acc Chem Res 37(10):798–804
76. Zimmerman SC, Zharov I, Wendland MS, Rakow NA, Suslick KS (2003) Molecular imprinting inside dendrimers. J Am Chem Soc 125(44):13504–13518
77. Slade CJ, Vulfson EN (1998) Induction of catalytic activity in proteins by lyophilization in the presence of a transition state analogue. Biotechnol Bioeng 57(2):211–215
78. Meng ZH, Yamazaki T, Sode K (2003) Enhancement of the catalytic activity of an artificial phosphotriesterase using a molecular imprinting technique. Biotechnol Lett 25(13):1075–1080
79. Beach IV, Shea KJ (1994) Designed catalysts. A synthetic network polymer that catalyzes the dehydrofluorhration of 4-fluoro-4-(p-nitrophenyl) butan-2-one. J Am Chem Soc 116(1):379–380
80. Muller R, Andersson LI, Mosbach K (1993) Molecularly imprinted polymers facilitating a beta-elimination reaction. Makromol. Chem Rapid Commun 14(10):637–641
81. Biffis A, Wulff G (2001) Molecular design of novel transition state analogues for molecular imprinting. New J Chem 25(12):1537–1542
82. Burri E, Ohm M, Daguenet C, Severin K (2005) Site-isolated porphyrin catalysts in imprinted polymers. Chem-A Euro J 11(17):5055–5061
83. Fireman-Shoresh S, Avnir D, Marx S (2003) General method for chiral imprinting of sol-gel thin films exhibiting enantioselectivity. Chem Mater 15(19):3607–3613
84. Visnjevski E, Yilmaz E, Bruggemanna O (2004) Catalyzing a cycloaddition with molecularly imprinted polymers obtained via immobilized templates. Appl Catal A 260(2):169–174
85. Yilmaz E, Haupt K, Mosbach K (2000) The use of immobilized templates – a new approach in molecular imprinting. Angew Chem Int Ed 39(12):2115–2118
86. Titirici MM, Hall AJ, Sellergren B (2002) Hierarchically imprinted stationary phases: mesoporous polymer beads containing surface-confined binding sites for adenine. Chem Mater 14(1):21–23
87. Nicholls IA, Rosengren JP (2002) Molecule selective surfaces. Bioseparat 10(3):301–305
88. Sode K, Ohta S, Yanai Y, Yamazaki T (2003) Construction of a molecular imprinting catalyst using target analogue template and its application for an amperometric fructosylamine sensor. Biosens Bioelectr 18(12):1485–1490
89. Sode K, Ishimura Tsugawa FW (2001) Screening and characterization of fructosylvaline utilizing marine microorganisms. Marine Biotechnol 3(2):126–132
90. Bruggemann O (2001) Chemical reaction engineering using molecularly imprinted polymeric catalysts. Anal Chim Act 435(11):197–207

Editor: Karel Dušek

Author Index Volumes 201–206

Author Index Volumes 1–100 see Volume 100
Author Index Volumes 101–200 see Volume 200

Anwander, R. see Fischbach, A.: Vol. 204, pp. 155–290.
Ayres, L. see Löwik D. W. P. M.: Vol. 202, pp. 19–52.

Boutevin, B., David, G. and *Boyer, C.*: Telechelic Oligomers and Macromonomers by Radical Techniques. Vol. 206, pp. 31–135
Boyer, C., see Boutevin B: Vol. 206, pp. 31–135

Csetneki, I., see Filipcsei G: Vol. 206, pp. 137–189

David, G., see Boutevin B: Vol. 206, pp. 31–135
Deming T. J.: Polypeptide and Polypeptide Hybrid Copolymer Synthesis via NCA Polymerization. Vol. 202, pp. 1–18.
Donnio, B. and *Guillon, D.*: Liquid Crystalline Dendrimers and Polypedes. Vol. 201, pp. 45–156.

Elisseeff, J. H. see Varghese, S.: Vol. 203, pp. 95–144.

Ferguson, J. S., see Gong B: Vol. 206, pp. 1–29
Filipcsei, G., Csetneki, I., Szilágyi, A. and *Zrínyi, M.*: Magnetic Field-Responsive Smart Polymer Composites. Vol. 206, pp. 137–189
Fischbach, A. and *Anwander, R.*: Rare-Earth Metals and Aluminum Getting Close in Ziegler-type Organometallics. Vol. 204, pp. 155–290.
Fischbach, C. and *Mooney, D. J.*: Polymeric Systems for Bioinspired Delivery of Angiogenic Molecules. Vol. 203, pp. 191–222.
Freier T.: Biopolyesters in Tissue Engineering Applications. Vol. 203, pp. 1–62.
Friebe, L., Nuyken, O. and *Obrecht, W.*: Neodymium Based Ziegler/Natta Catalysts and their Application in Diene Polymerization. Vol. 204, pp. 1–154.

García A. J.: Interfaces to Control Cell-Biomaterial Adhesive Interactions. Vol. 203, pp. 171–190.
Gong, B., Sanford, AR. and *Ferguson, JS.*: Enforced Folding of Unnatural Oligomers: Creating Hollow Helices with Nanosized Pores. Vol. 206, pp. 1–29
Guillon, D. see Donnio, B.: Vol. 201, pp. 45–156.

Harada, A., Hashidzume, A. and *Takashima, Y.*: Cyclodextrin-Based Supramolecular Polymers. Vol. 201, pp. 1–44.
Hashidzume, A. see Harada, A.: Vol. 201, pp. 1–44.

Heinze, T., Liebert, T., Heublein, B. and *Hornig, S.*: Functional Polymers Based on Dextran. Vol. 205, pp. 199–291.
Heßler, N. see Klemm, D.: Vol. 205, pp. 57–104.
Van Hest J. C. M. see Löwik D. W. P. M.: Vol. 202, pp. 19–52.
Heublein, B. see Heinze, T.: Vol. 205, pp. 199–291.
Hornig, S. see Heinze, T.: Vol. 205, pp. 199–291.
Hornung, M. see Klemm, D.: Vol. 205, pp. 57–104.

Jaeger, W. see Kudaibergenov, S.: Vol. 201, pp. 157–224.
Janowski, B. see Pielichowski, K.: Vol. 201, pp. 225–296.

Kataoka, K. see Osada, K.: Vol. 202, pp. 113–154.
Klemm, D., Schumann, D., Kramer, F., Heßler, N., Hornung, M., Schmauder H.-P. and *Marsch, S.*: Nanocelluloses as Innovative Polymers in Research and Application. Vol. 205, pp. 57–104.
Klok H.-A. and *Lecommandoux, S.*: Solid-State Structure, Organization and Properties of Peptide—Synthetic Hybrid Block Copolymers. Vol. 202, pp. 75–112.
Kosma, P. see Potthast, A.: Vol. 205, pp. 151–198.
Kosma, P. see Rosenau, T.: Vol. 205, pp. 105–149.
Kramer, F. see Klemm, D.: Vol. 205, pp. 57–104.
Kudaibergenov, S., Jaeger, W. and *Laschewsky, A.*: Polymeric Betaines: Synthesis, Characterization, and Application. Vol. 201, pp. 157–224.

Laschewsky, A. see Kudaibergenov, S.: Vol. 201, pp. 157–224.
Lecommandoux, S. see Klok H.-A.: Vol. 202, pp. 75–112.
Li, S., see Li W: Vol. 206, pp. 191–210
Li, W. and *Li, S.*: Molecular Imprinting: A Versatile Tool for Separation, Sensors and Catalysis. Vol. 206, pp. 191–210
Liebert, T. see Heinze, T.: Vol. 205, pp. 199–291.
Löwik, D. W. P. M., Ayres, L., Smeenk, J. M., Van Hest J. C. M.: Synthesis of Bio-Inspired Hybrid Polymers Using Peptide Synthesis and Protein Engineering. Vol. 202, pp. 19–52.

Marsch, S. see Klemm, D.: Vol. 205, pp. 57–104.
Mooney, D. J. see Fischbach, C.: Vol. 203, pp. 191–222.

Nishio Y.: Material Functionalization of Cellulose and Related Polysaccharides via Diverse Microcompositions. Vol. 205, pp. 1–55.
Njuguna, J. see Pielichowski, K.: Vol. 201, pp. 225–296.
Nuyken, O. see Friebe, L.: Vol. 204, pp. 1–154.

Obrecht, W. see Friebe, L.: Vol. 204, pp. 1–154.
Osada, K. and *Kataoka, K.*: Drug and Gene Delivery Based on Supramolecular Assembly of PEG-Polypeptide Hybrid Block Copolymers. Vol. 202, pp. 113–154.

Pielichowski, J. see Pielichowski, K.: Vol. 201, pp. 225–296.
Pielichowski, K., Njuguna, J., Janowski, B. and *Pielichowski, J.*: Polyhedral Oligomeric Silsesquioxanes (POSS)-Containing Nanohybrid Polymers. Vol. 201, pp. 225–296.
Pompe, T. see Werner, C.: Vol. 203, pp. 63–94.
Potthast, A., Rosenau, T. and *Kosma, P.*: Analysis of Oxidized Functionalities in Cellulose. Vol. 205, pp. 151–198.
Potthast, A. see Rosenau, T.: Vol. 205, pp. 105–149.

Rosenau, T., Potthast, A. and *Kosma, P.*: Trapping of Reactive Intermediates to Study Reaction Mechanisms in Cellulose Chemistry. Vol. 205, pp. 105–149.
Rosenau, T. see Potthast, A.: Vol. 205, pp. 151–198.

Salchert, K. see Werner, C.: Vol. 203, pp. 63–94.
Sanford, A. R., see Gong B: Vol. 206, pp. 1–29
Schlaad H.: Solution Properties of Polypeptide-based Copolymers. Vol. 202, pp. 53–74.
Schmauder H.-P. see Klemm, D.: Vol. 205, pp. 57–104.
Schumann, D. see Klemm, D.: Vol. 205, pp. 57–104.
Smeenk, J. M. see Löwik D. W. P. M.: Vol. 202, pp. 19–52.
Szilágyi, A., see Filipcsei G: Vol. 206, pp. 137–189

Takashima, Y. see Harada, A.: Vol. 201, pp. 1–44.

Varghese, S. and *Elisseeff, J. H.*: Hydrogels for Musculoskeletal Tissue Engineering. Vol. 203, pp. 95–144.

Werner, C., Pompe, T. and *Salchert, K.*: Modulating Extracellular Matrix at Interfaces of Polymeric Materials. Vol. 203, pp. 63–94.

Zhang, S. see Zhao, X.: Vol. 203, pp. 145–170.
Zhao, X. and *Zhang, S.*: Self-Assembling Nanopeptides Become a New Type of Biomaterial. Vol. 203, pp. 145–170.
Zrínyi, M., see Filipcsei G: Vol. 206, pp. 137–189

Subject Index

Acid chlorides   13
Acrylates   44
Acrylic double bonds   91
Addition–fragmentation, catalytic chain transfer   105
–, telechelic oligomers   47
Addition–fragmentation chain transfer, reversible   72
Alkyl (meth)acrylate, thiolate   38
Amines, aromatic   13
Anisotropic elastomers   137, 145
Anisotropic mechanical behavior   155
APTME   194
Atom transfer radical coupling (ATRC)   69
Atrazines   200
ATRP, macromonomers   110
–, telechelic oligomers   58
4,4′-Azobis(4-cyanovaleric acid) (ACVA)   40

BHEBT   74
Bromo polystyrene oligomers   71
n-Butyl methacrylate, telomerization   40

Ca-alginate   149
Catalysis materials, alternative   201
Catalysis, molecular imprinting   191
Catalytic chain transfer   52, 106
Cavity size   12
Chain transfer agents, addition–fragmentation   47
Chain-end chemical modification, telechelic oligomers   39
Chiral stationary phases (CSPs)   195
Chloramphenicol diacetate   199
Chloramphenicol-methyl red   199
Compressive force   162
Conventional radical polymerizations   31
Coupling   69

–, acid chlorides/aromatic amines   13
Crescent oligoamides   7

Dead-end polymerization (DEP)   34
–, telechelic oligomers   41
Diazepam   197
Diels–Alder cycloadditions   203
Diethylaminoethyl methacrylate (DEAEM)   200
Diiodo poly(vinyl chloride)   88
1,6-Dimethacrylate hexane, 2-mercaptoethanol   39
Dithiols, radical addition   38
Double bonds   91, 95

Elastic modulus, uniform magnetic field   159
Elastomers   143
Ethylene glycol dimethacrylate (EDMA)   192, 194

Ferrofluids   142
Ferrogels   137, 140
–, magnetic properties   150
Fluoro-type monomers   46
Foldamer   1
Fructosylamine dehydrogenase   205

Glutaraldehyde   144
Guanidinium ion, receptors   17

Helices, nanoporous   1
Hexamethyltriethylenetetramine (HMDETA)   110
Hollow helices, nanosized pores   1
Hydrogen bond, three-center   3

Inifer/iniferter systems   53
Interpenetrated network (IPN)   149

Subject Index

Iodine transfer polymerization, telechelic oligomers 86

Latexes, magnetic 137

Macrocyclization, folding-assisted 13
Macromonomers 31
–, polycondensable groups 96, 104
–, polymerizable double bond 98, 110
–, radical techniques 90
–, telomerization 96
MADIX 72
Magnetic composites 137
–, elastic properties, unidirectional compression 153
Magnetic elastomers, uniform field 163
Magnetic field-induced deformation 166
Magnetic field-sensitive polymer gels 137, 140
Magnetic gel beads 147
Magnetic latexes 137
Magnetic particles, swelling behavior 178
Magnetic polymer composites/gels 141, 143
Magnetoelast 137, 154
Maleimide, terpyridine-functionalized 85
Membranes, electro-resistance 200
–, smart 137
Methyl orange, selective adsorbent 192
MMA 108
–, lignin-terminated 96
Molecular imprinting 191
Molecularly imprinted polymers (MIPs) 191
Monohydroxy oligomethylmethacrylates 77
Monolith magnetic polymers 143
Mooney–Rivlin representation 154

Naproxen 197
NIPA 149
NIPAM 102
Nitroxide-mediated polymerization, macromonomers 118
–, telechelic oligomers 79
Nitroxides, living free-radical polymerization 79
Non-homogeneous deformation 173
Nucleophilic addition, telechelic oligomers 38

Oligoamides, aromatic 1
–, backbone-rigidified, tunable cavity 3
–, macrocycles, shape-persistent 15
Oligoanthranilimides 23
Oligomers, unnatural, enforced folding 1
Oligomethacrylates, hydroxy-telechelic 40
Oligomethylmethacrylates, monohydroxy 77
Oligo(m-phenylene ethynylenes), enforced folding 20
Oxidative cleavage 57

PDMS gels, magnetic particles 178
Pentamethyldiethylenetriamine (PMDETA)
PMMA 78, 108
PNIPA 178
–, millimeter-sized 148
–, thermosensitive gel beads 148
Poly(n-butyl acrylate), carboxy-functional 76
Poly(dimethylsiloxane) elastomers, magnetic 144
Poly(N-isopropylacrylamide) (PNIPA) 148
Poly(vinyl alcohol) gels, magnetite-loaded 143, 161
Polybutadiene, hydroxy-telechelic 84
Polycondensable groups, macromonomers 115
Polymer gels, magnetic field-sensitive 137
–, microspheres, stimuli-responsive 142
Polymers, molecularly imprinted (MIPs) 191
Polymethacrylate telomers 39
Polystyrene, terpyridine-telechelic 85
Polystyrene latex, magnetic 147
–/PNIPA gel shell 149
Polyurethanes 36
mPVA gels/mPDMS networks 156

Radical addition 89
–, telechelic oligomers 36
Radical coupling 69, 89
Radical polymerization, controlled/living 31
RAFT 72

Self-assembly, intermolecular interaction 193

Sensors, molecular imprinting   191
–, substrate-selective   198
Separation, molecular imprinting   191
Separation materials, tailor-made   195
Smart membranes   137
Stimuli-responsive polymer gel microspheres   142
Stress–strain dependence   137
Styrene   42, 110
–/divinylbenzene (DVB)   192
Styrenic double bonds   91
Superparamagnetism   150

Telechelic oligomers   31
–, radical techniques   35
–, telomerization   36
Telomerization, telechelic oligomers   36
TEMPO   69, 81
Temporary reinforcement   137
Terpyridine ruthenium   85

Tetrachlorobutyl acetate   37
Tetrasulfonamide macrocycles, cavity-containing   22
Tetraurea macrocycles, folding-assisted aromatic   21
Theophylline   197
Thiamphenicol (TAM)   199
Thioester modification   76
Transition state analog   202
Trimethylolpropane trimethacrylate (TRIM)   194
Trithiocarbonates   74
Trithioester transfer agent   74

Vibration, shock absorbers   137
Vinyl chloroacetate   110
Vinylbenzyltrimethylammonium chloride   74
4-Vinylpyridine   197

Xanthates   72

Printing: Krips bv, Meppel
Binding: Stürtz, Würzburg